Combinatorics and Renormalization in Quantum Field Theory

FRONTIERS IN PHYSICS

DAVID PINES, Editor

FRONTIERS IN PHYSICS

FRONTIERS IN PHYSICS

DAVID PINES, Editor

Combinatorics and Renormalization in Quantum Field Theory

E. R. Caianiello

University of Salerno and Laboratory of Cybernetics, Naples

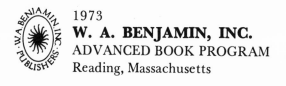

1973

W. A. BENJAMIN, INC.

ADVANCED BOOK PROGRAM

Reading, Massachusetts

London · Amsterdam · Don Mills, Ontario · Sydney · Tokyo

6666-0397

PHYSICS

Library of Congress Cataloging in Publication Data

Caianiello, Eduardo R   1921–
    Combinatorics and renormalization in quantum field
theory.

    (Frontiers in physics)
    Includes bibliographical references.
    1. Quantum field theory.   2.   Combinatorial
analysis.   3.   Renormalization (Physics)   I.   Title.
II.   Series.
QC174.45.C34   530.1'43   73–4610

ISBN 0-8053-1646-9
ISBN 0-8053-1645-0 (pbk.)

Contents

PART II.   EQUATIONS FOR PROPAGATORS AND PERTURBATIVE
           EXPANSIONS

PART III.   REGULARIZATION, RENORMALIZATION, AND MASS
            EQUATIONS

EDITOR'S FOREWORD

The problem of communicating in a coherent fashion the recent developments in the most exciting and active fields of physics seems particularly pressing today. The enormous growth in the number of physicists has tended to make the familiar channels of communication considerably less effective. It has become increasingly difficult for experts in a given field to keep up with the current literature; the novice can only be confused. What is needed is both a consistent account of a field and the presentation of a definite "point of view" concerning it. Formal monographs cannot meet such a need in a rapidly developing field, and, perhaps more important, the review article seems to have fallen into disfavor. Indeed, it would seem that the people most actively engaged in developing a given field are the people least likely to write at length about it.

"Frontiers in Physics" has been conceived in an effort to improve the situation in several ways. First, to take advantage of the fact that the leading physicists today frequently give a series of lectures, a graduate seminar, or a graduate course in their special fields of interest. Such lectures serve to summarize the present status of a rapidly developing field and may well constitute the only coherent account available at the time. Often, notes on lectures exist (prepared by the lecturer himself, by graduate students, or by postdoctoral fellows) and have been distributed in mimeographed form on a limited basis. One of the principal purposes of the "Frontiers in Physics" series is to make such notes available to a wider audience of physicists.

It should be emphasized that lecture notes are necessarily rough and informal, both in style and content, and those in the series will prove no exception. This is as it should be. The point of the series is to offer new, rapid, more informal, and, it is hoped, more effective ways for physicists to teach one another. The point is lost if only elegant notes qualify.

A second way to improve communication in very active fields of physics is by the publication of collections of reprints of recent articles. Such collections are themselves useful to people working in the field. The value of the reprints would, however, seem much enhanced if the collection would be accompanied by an introduction of moderate length, which would serve to tie the collection together and, necessarily, constitute a brief survey of the present status of the field. Again, it is appropriate that such an introduction be informal, in keeping with the active character of the field.

A third possibility for the series might be called an informal monograph, to connote the fact that it represents an intermediate step between lecture notes and formal monographs. It would offer the author an opportunity to present his views of a field that has developed to the point at which a summation might prove extraordinarily fruitful, but for which a formal monograph might not be feasible or desirable.

Fourth, there are the contemporary classics—papers or lectures which constitute a particularly valuable approach to the teaching and learning of physics today. Here one thinks of fields that lie at the heart of much of present-day research, but whose essentials are by now well understood, such as quantum electrodynamics or magnetic resonance. In such fields some of the best pedagogical material is not readily available, either because it consists of papers long out of print or lectures that have never been published.

"Frontiers in Physics" is designed to be flexible in editorial format. Authors are encouraged to use as many of the foregoing approaches as seem desirable for the project at hand. The publishing format for the series is in keeping with its intentions. Photo-offset printing is used throughout, and the books are paperbound, in order to speed publication and reduce costs. It is hoped that the books will thereby be within the financial reach of graduate students in this country and abroad.

Finally, because the series represents something of an experiment on the part of the editor and the publisher, suggestions from interested readers as to format, contributors, and contributions will be most welcome.

DAVID PINES

Urbana, Illinois
August 1961

Preface

This book is an outgrowth of a series of lectures delivered at the Department of Theoretical Physics of Imperial College, London, in 1971. The object of the lecture course was to give a comprehensive view of work done by my collaborators and myself, over a period of almost 20 years, on various aspects of field theory. Our approach is based upon a systematic investigation of combinatoric and analytic properties, kept as much apart as possible.

Part I contains a description of combinatoric methods; all results which have proved relevant in our later work are reported. The first and third chapters cover material that was presented in a course at Princeton University in 1956; the second chapter, a digression from the main subject, is intended to give some perspective on relevant applications of this formalism to subjects other than quantum field theory. (Its main application, which is found in the study of the Ising model, is not reported in Chapter 2, however, since excellent literature on the subject is plentiful.) In conclusion, it is shown that combinatorics is the simple and natural means of exploiting the remarkable properties of Wick products, so that all expansions and formulas of field theory can be obtained, without diagrammatic expansions, with the same ease as, and as a generalization of, the standard formal theory of Fredholm equations.

Part II also deals with the formal theory, which is defined in x space by means of the equations that connect propagators (or Green's functions) among themselves in various ways. Unitarity, gauge invariance, and infrared divergences are studied with the methods described in Part I, in order to achieve uniformity of treatment as well as economy and simplicity. Of course, nothing new is learned, although proofs that often exceed 50 pages in length gain tremendously in concision: for instance, the linked-cluster expansion, as it was called when rediscovered in the context of many-body theory, took only a few lines of proof. Some consideration is also given to the convergence and divergence of regularized expansions and to solutions of simple models.

Part III is given over to analytic problems, which arise as soon as regularizations are removed and equations and expansions become undefined or infinite wherever a product of distributions misbehaves. This is the subject of renormalization, which is treated in a general way by asking: Under what conditions does a regularization lead to results independent of the regularizing procedure, so as not to affect the physics? Again, the main differences from the classic treatments lie only in the use of combinatorics, which permits a very great condensation of the demonstrations, and in the fact that renormalization is applied directly to the equations that define the propagators, perturbative expansions coming out renormalized as a consequence. It is also shown that, however different any two renormalization procedures may appear, they can always be related to each other and to the same physics by means of simple steps, which arise naturally from the consideration of infinitesimal renormalizations. This point is discussed further in the first section of Chapter 9, which the reader should consider as part of this Preface.

The last chapter examines the behavior of some very simple approximations and solvable models, when the renormalization methods described in the text are applied and all expressions computed exactly. In all of them the physical mass appears as the solution of a mass equation, which can have more than one solution; the relation of this behavior to inequivalent representations is discussed briefly. This issue, as well as that of the *existence* of solutions, which is definitely avoided in our treatment, will require further work, for which I believe this formalism presents an alternative approach to that offered by constructive field theory.

Proofs are mostly omitted, references being readily available. This may not be a disadvantage, since our main result is perhaps that the knowledge of the techniques required for them becomes unnecessary for the use of the theory. It should also be added that the main emphasis throughout is on keeping the exposition as elementary as possible: at the cost of seeming naïve, but not of rigor, the plainest mathematical language and formalisms have always been preferred, even when more sophisticated ones are current.

It is my duty and pleasure to express my warmest thanks to Prof. M. Marinaro for her invaluable assistance with both the physics and the elaboration of this material; to Professors T. W. B. Kibble, P. T. Matthews, and A. Salam for their warm hospitality at Imperial College; and to Drs. G. Leibbrandt and R. M. Williams for their generous help in writing these notes and for many fruitful discussions.

E. R. CAIANIELLO

Combinatorics and Renormalization in

Quantum Field Theory

Combinatorics and Renormalization in Quantum Field Theory

Part I

Combinatoric Methods

Chapter 1

Basic Combinatoric Tools: Pfaffians

1. INTRODUCTION

In this chapter we introduce the basic algorithms that are necessary to cast any perturbative expansion of quantum field or many-body physics into a form as simple as that of the classic Fredholm theory (which turns out to be a particular instance among those covered by this formalism). Fermi and Bose statistics require, respectively, the definition and study of *pfaffians* and *determinants,* and of *hafnians* and *permanents* [1]: any expectation value of any product of free Fermi fields is in fact a pfaffian (which often, as we shall see, reduces to a determinant, e.g., in the Fredholm case and in electrodynamics); of Bose fields, a hafnian (which often reduces to a permanent). Tremendous simplifications are trivially seen to occur if we start with some preliminary (and quite elementary) mathematical considerations concerning these objects and their main properties; this we do here, with special emphasis on pfaffians and their relation to determinants (the corresponding statements on hafnians and permanents, when true, will be derived from the former with only trivial changes of notation).

The uses of this formalism are many, in various domains of science; were it better known, it could no doubt find further profitable applications. To stimulate thought in these directions, and to give a concrete familiarity with it, is the only object of the second chapter, which can otherwise be omitted without loss as regards our main topic.

The third chapter is mostly dedicated to the enunciation of some simple, though little-known, expansion theorems which hold, *mutatis mutandis,* for all four of the algorithms just mentioned. The use of these theorems suffices, as we shall indicate in Chapter 9, to automatize entirely all proofs concerning renormalization, counterterms, and the like, obviating the need for painstaking graph-by-graph analyses; it thus helps greatly in developing, as is our intent in this book, a renormalization theory both of the equations for Green's functions and of their perturbative expansions (which is neither invalidated by lack of convergence of the latter, nor depends on the choice of particular solution or approximation techniques). The last section of Chapter 3 gives, as the conclusion of Part I, the explicit perturbative expansion for electrodynamics (it can be written on inspection with equal ease for any other theory), which evidences the role of the combinatoric tools here described.

2. ANTISYMMETRIC DETERMINANTS

We shall often denote the determinant of a given (square) matrix A

$$
\det A = \begin{vmatrix} a_{11} & \cdot & \cdots & a_{1n} \\ a_{21} & a_{22} & \cdots & \cdot \\ \cdot & \cdot & \cdots & \cdot \\ \cdot & \cdot & \cdots & \cdot \\ \cdot & \cdot & \cdots & \cdot \\ a_{n1} & \cdot & \cdots & a_{nn} \end{vmatrix} \tag{1.1}
$$

with the shortened notation, due to Cayley,

$$
\det A = \begin{pmatrix} 1 \cdots n \\ 1 \cdots n \end{pmatrix} \tag{1.2}
$$

where the first line in (1.2) lists the row indices, while the second line gives the column indices, of the determinant (1.1). Thus, if A is of order 3,

$$
\begin{vmatrix} a_{11} & a_{12} & a_{13} \\ a_{21} & a_{22} & a_{23} \\ a_{31} & a_{32} & a_{33} \end{vmatrix} \stackrel{\text{def}}{=\!=\!=} \begin{pmatrix} 1 & 2 & 3 \\ 1 & 2 & 3 \end{pmatrix}.
$$

The notation (1.2) is convenient when expansions and combinatoric manipulations are involved, which is, in general, when the explicit values of the elements a_{hk}, which are tabulated in (1.1), are not required. The formal gains are, in the present context, comparable to those offered by tensor versus fully expanded notation.

An *antisymmetric* (or skew-symmetric) determinant has

$$
a_{hk} = -a_{kh}, \qquad h, k = 1, \ldots, n.
$$

For example,

$$
\begin{vmatrix}
0 & a_{12} & a_{13} & a_{14} \\
-a_{12} & 0 & a_{23} & a_{24} \\
\cdot & \cdot & 0 & \cdot \\
-a_{14} & \cdot & \cdot & 0
\end{vmatrix}
= \begin{pmatrix} 1 & 2 & 3 & 4 \\ 1 & 2 & 3 & 4 \end{pmatrix}
\tag{1.3}
$$

The value of an antisymmetric determinant depends on its *order:*

   (i)   If the order is *odd*, det $A = 0$.
   (ii) If the order is *even*, det A is a perfect square.

For example, the determinant (1.3) may be written as

$$
\det A = \begin{pmatrix} 1 & 2 & 3 & 4 \\ 1 & 2 & 3 & 4 \end{pmatrix} = (a_{12}a_{34} - a_{13}a_{24} + a_{14}a_{23})^2
$$

$$
= (\mathrm{pf}\, A)^2 = [(1\ \ 2\ \ 3\ \ 4)]^2
$$

if we write

$$
\mathrm{pf}\, A = a_{12}a_{34} - a_{13}a_{24} + a_{14}a_{23} = (1\ \ 2\ \ 3\ \ 4).
$$

This expression is a particular case of the general theorem, valid for any antisymmetric matrix A of *even* order [2]:

$$
\det A = \begin{pmatrix} 1 & \cdots & 2m \\ 1 & \cdots & 2m \end{pmatrix} = [\mathrm{pf}\, A]^2 = [(1 \cdots 2m)]^2
$$

where the pfaffian connected to the matrix A, or equivalently to its elements above the main diagonal, is defined in the next section (this algorithm was introduced by Jacobi, while studying the Pfaff problem).

3. PFAFFIANS

It is convenient to represent a pfaffian by the triangular array

$$
\mathrm{pf}\, A = (1 \cdots 2m) = \begin{vmatrix} (1\ 2) & (1\ 3) & \cdots & (1\ 2m) \\ & (2\ 3) & \cdots & (2\ 2m) \\ & & \cdot & \\ & & \cdot & \\ & & \cdot & \\ & & & (2m-1\ 2m) \end{vmatrix}
\tag{1.4}
$$

where (hk) denotes the element a_{hk} with $h < k$. A pfaffian can be expanded by the elements of one of its *lines*: "line h" is that line in the triangular array (1.4) which contains all the elements carrying the index regardless whether h is the first or second index. For example, "line 2" i

$$(1\ 2\ 3\ 4) = \begin{vmatrix} (1\ 2) & (1\ 3) & (1\ 4) \\ & (2\ 3) & (2\ 4) \\ & & (3\ 4) \end{vmatrix}$$

contains the elements (1 2), (2 3) and (2 4). Deletion of lines h and k yields another pfaffian, called the *minor* of (hk), in analogy with the theory of determinants. The adjoint of an element (hk) is its minor with sign $(-1)^{h+k+1}$. Thus if we expand (1 2 3 4) in terms of line 2, we get (cf. (1.25))

$$(1\ 2\ 3\ 4) = (1\ 2)(3\ 4) + (2\ 3)(1\ 4) - (2\ 4)(1\ 3).$$

The *expansion rule for pfaffians* can be summarized as follows:

$$(1 \cdots 2m) = \sum_{h=2}^{2m} (-1)^h (1h)(2, \ldots, h-1, h+1, \ldots, 2m) \qquad (1.5)$$

(Expansion in terms of line 1)

$$= \Sigma' \, (-1)^P (i_1 i_2)(i_3 i_4) \cdots (i_{2m-1} i_{2m}),$$

where Σ' means summation over all permutations $i_1 i_2 \cdots i_{2m}$ of $(1 \cdots 2m)$ which satisfy the inequalities

$$i_1 < i_2; \quad i_3 < i_4; \ldots; i_{2m-1} < i_{2m}; \qquad i_1 < i_3 < i_5 \cdots < i_{2m-1}$$

In words, rule (1.5) states that the pfaffian $(1 \cdots 2m)$ is equal to the sum of the products of each element (hk) of a prefixed line by its adjoin

Example.   Expansion of the pfaffian

$$(1\ 2\ 3\ 4\ 5\ 6) = \begin{vmatrix} (1\ 2) & (1\ 3) & \cdots & (1\ 6) \\ & (2\ 3) & \cdots & (2\ 6) \\ & & \cdot & \cdot \\ & & \cdot & \cdot \\ & & \cdot & \cdot \\ & & & (5\ 6) \end{vmatrix}$$

in terms of line 1 yields

$$(1\ 2\ 3\ 4\ 5\ 6) = (-1)^{1+2+1}(1\ 2)(3\ 4\ 5\ 6) + (-1)^{1+3+1}(1\ 3)(2\ 4\ 5\ 6)$$
$$+ (-1)^{1+4+1}(1\ 4)(2\ 3\ 5\ 6) + (-1)^{1+5+1}(1\ 5)(2\ 3\ 4\ 6)$$
$$+ (-1)^{1+6+1}(1\ 6)(2\ 3\ 4\ 5)$$
$$= (1\ 2)(3\ 4\ 5\ 6) - (1\ 3)(2\ 4\ 5\ 6) + (1\ 4)(2\ 3\ 5\ 6)$$
$$- (1\ 5)(2\ 3\ 4\ 6) + (1\ 6)(2\ 3\ 4\ 5).$$

4. LINEAR AND QUADRATIC FORMS AND THE GRASSMANN PRODUCT

Let $x^1, \ldots, x^N$ be a finite or infinite set of elements satisfying the multiplication law

$$x^h \wedge x^k = -x^k \wedge x^h \qquad (x^h \wedge x^k = 0). \tag{1.6}$$

The product (1.6) is a *Grassmann (or exterior) product,* and the symbol $\wedge$ denotes multiplication in a Grassmann tensor algebra (also called an exterior algebra) [3, 4]. Consider now all linear forms of the type

$$\omega_h = \sum_{k=1}^{n} a_{hk} x^k$$

where the ω are defined over a Grassmann algebra generated by the elements x^h, while the a_{hk}'s are arbitrary complex numbers. The product of two linear forms is given by

$$\omega_1 \wedge \omega_2 = \sum_{h \neq k} a_{1h} a_{2k} x^h \wedge x^k$$

$$= \sum_{h < k} \begin{pmatrix} 1 & 2 \\ h & k \end{pmatrix} x^h \wedge x^k; \tag{1.7}$$

$\begin{pmatrix} 1 & 2 \\ h & k \end{pmatrix}$ is the determinantal notation introduced in Section 2. The Grassmann product of m linear forms ω_h reads

$$\omega_1 \wedge \omega_2 \wedge \cdots \wedge \omega_m = \sum_{h_1 < h_2 < \cdots < h_m} \begin{pmatrix} 1 & \cdots & m \\ h_1 & \cdots & h_m \end{pmatrix} x^{h_1} \wedge \cdots \wedge x^{h_m}.$$
$$\tag{1.8}$$

For $m = N$ the (finite) product (1.8) reduces to

$$\omega_1 \wedge \omega_2 \wedge \cdots \wedge \omega_N = \begin{pmatrix} 1 \cdots N \\ 1 \cdots N \end{pmatrix} x^1 \wedge x^2 \wedge x^3 \wedge \cdots \wedge x^N. \quad (1.9)$$

The foregoing considerations on linear forms can also be applied to *quadratic* forms:

$$\Omega = \frac{1}{2} \sum_{h,k} \alpha_{hk} x^h \wedge x^k, \qquad \alpha_{hk} = -\alpha_{kh} \quad (1.10)$$

$$= \sum_{h<k} \alpha_{hk} x^h \wedge x^k.$$

Considering powers of the *same* quadratic form Ω, we have

$$\Omega_\wedge^2 \equiv \Omega \wedge \Omega = \sum_{h_1<h_2} \sum_{h_3<h_4} \alpha_{h_1 h_2} \alpha_{h_3 h_4} x^{h_1} \wedge x^{h_2} \wedge x^{h_3} \wedge x^{h_4}$$

$$= 2! \sum_{h_1<h_2<h_3<h_4} (h_1 h_2 h_3 h_4) x^{h_1} \wedge x^{h_2} \wedge x^{h_3} \wedge x^{h_4}, \quad (1.11)$$

with the pfaffian $(h_1 h_2 h_3 h_4)$ given by

$$(h_1 h_2 h_3 h_4) = \alpha_{h_1 h_2} \alpha_{h_3 h_4} - \alpha_{h_1 h_3} \alpha_{h_2 h_4} + \alpha_{h_1 h_4} \alpha_{h_2 h_3}.$$

It can easily be deduced from (1.11) that the mth power of the quadratic form Ω is

$$\Omega_\wedge^m = m!(1 \cdots 2m) x^1 \wedge x^2 \wedge \cdots \wedge x^{2m}.$$

5. RELATION BETWEEN GRASSMANN'S ALGEBRA AND CLIFFORD'S ALGEBRA

Let $x^1, x^2, \ldots, x^n$ be the "same set of elements with another multiplication law":

$$x^h \wedge x^k = -x^k \wedge x^h + 2\delta_{hk} \qquad (x^h \wedge x^k = 1). \quad (1.12)$$

The product (1.12) is a Clifford product and δ_{hk} is the Kronecker symbol defined in the usual way; the symbol $\wedge$ denotes multiplication in a *Clifford algebra* [4]. If we consider all linear forms

$$\omega_h = \sum_i a_{hi} x^i, \quad (1.13)$$

where the a_{hi}'s are arbitrary complex numbers, then the ω's span a Clifford algebra, for which

$$\omega_h \wedge \omega_k = -\omega_k \wedge \omega_h + 2(hk), \tag{1.14}$$

with $(hk) = \Sigma_i \, \alpha_{hi} \alpha_{ki}$. The products

$$\Omega_{h_1 h_2 \cdots h_k} = \omega_{h_1} \wedge \omega_{h_2} \wedge \cdots \wedge \omega_{h_k} \tag{1.15}$$

are the "simple" elements of the corresponding tensor *Clifford algebra* which is spanned in general by all linear combinations of the "simple" elements (1.15).

We are particularly interested in the totally antisymmetric tensors $\mathscr{A}^{(k)}$ of the Clifford algebra, generated by all elements (1.15). Taking $k = 2$, for example, we obtain the tensor

$$\mathscr{A}_{hk}^{(2)} = \frac{1}{2!} \, \{\omega_h \wedge \omega_k - \omega_k \wedge \omega_h\}, \tag{1.16}$$

and using relation (1.14) we get

$$\omega_h \wedge \omega_k = \mathscr{A}_{hk}^{(2)} + (hk). \tag{1.17}$$

In order to establish a relation between the Grassmann product and the Clifford product we employ definitions (1.6) and (1.7) and set

$$\mathscr{A}_{hk}^{(2)} = \omega_h \wedge \omega_k = -\omega_k \wedge \omega_h. \tag{1.18}$$

It then follows from Eqs. (1.17) and (1.18) that

$$\omega_h \wedge \omega_k = \omega_h \wedge \omega_k + (hk). \tag{1.19}$$

The fact that (1.19) can be expressed by means of (1.16), or that (1.6) can be expressed by means of (1.12), will permit us to establish a relation of the utmost importance between Grassmann and Clifford algebras and products: Either type of product can be expressed as a linear combination of products of the other type. It is important to remark that the *totally antisymmetric simple tensors of the Clifford algebra can be identified with the corresponding simple tensors of the Grassmann algebra.* Thus

$$\mathscr{A}_{h_1 h_2 \cdots h_k}^{(k)} = \frac{1}{k!} \sum_P (-1)^P \omega_{h_{i_1}} \wedge \omega_{h_{i_2}} \wedge \cdots \wedge \omega_{h_{i_k}}$$

$$= \omega_{h_1} \wedge \omega_{h_2} \wedge \cdots \wedge \omega_{h_k}, \tag{1.20}$$

so that once the Clifford product is defined, the Grassmann product rule becomes defined as a consequence of it; P denotes any permutation $h_{i_1} \cdots h_{i_k}$ of $h_1 \cdots h_k$ of parity P. The proof of (1.20) follows immediately from Eqs. (1.6) and (1.12). We now state without proof the fundamental theorem connecting Grassmann and Clifford algebras [4].

Theorem

$$\omega_{h_1} \wedge \omega_{h_2} \wedge \cdots \wedge \omega_{h_k} = \sum_{r=0}^{[k/2]} \sum_{C_r} (-1)^{P_k} (i_1 i_2 \cdots i_{2r})$$

$$\times \omega_{j_1} \wedge \omega_{j_2} \wedge \cdots \wedge \omega_{j_{k-2r}}, \qquad (1.21)$$

where $[k/2]$ is the maximum integer contained in $k/2$; $i_1 < i_2 < \cdots < i_{2r}$, $j_1 < j_2 < \cdots < j_{k-2r}$ is a combination C_r of $h_1 h_2 \cdots h_k$ (here i_s, j_t are short forms for h_{i_s}, h_{j_t}). The latter establishes the "natural order"; $i_s < i_t$ means that i_s precedes i_t in this order, that is, $s < t$; P is the parity of $i_1 i_2 \cdots i_{2r}$ and $j_1 j_2 \cdots j_{k-2r}$ with respect to the natural order; $\sum_{C_r}$ denotes summation over all possible combinations C_r; $(i_1 \cdots i_{2r})$ is the pfaffian of elements defined by

$$(hk) = (kh) = \sum_i \alpha_{hi} \alpha_{ki}.$$

Example. We shall illustrate the foregoing theorem for the two linear forms ω_1 and ω_2. The Clifford product $\omega_1 \wedge \omega_2$ can then be written as

$$\omega_1 \wedge \omega_2 = (1\ 2) + \omega_1 \wedge \omega_2 \qquad (1.22)$$

where $(1\ 2) = \sum_i \alpha_{1i} \alpha_{2i}$.

6. FURTHER PROPERTIES OF PFAFFIANS

A. Symmetry Properties

We have learned that the pfaffian $(1 \cdots 2n)$ may be written as a triangular array:

$$(1 \cdots 2n) = \begin{vmatrix} (1\ 2) & (1\ 3) & \cdots & (1\ 2n) \\ & (2\ 3) & \cdots & (2\ 2n) \\ & & \ddots & \vdots \\ & & & (2n-1\ 2n) \end{vmatrix} \qquad (1.23)$$

where (hk) comes from the element a_{hk} of an antisymmetric determinant such that

$$a_{hk} = (hk) = -a_{kh} \qquad (h < k).$$

For example, $a_{13} = +(1\ 3)$, while $a_{31} = -(1\ 3)$. Our aim is to study the behavior of the pfaffian (1.23) under the interchange of any two lines, or under the permutation of several lines. (We recall that, in the case of determinants, the single interchange of any two columns or any two rows introduces a minus sign.) We state the following rule for permuting lines in a pfaffian.

Rule. (i) Write the pfaffian with elements a_{hk} instead of (hk) (it makes no difference in (1.23)).

(ii) Then:

$$(h_1 \cdots h_{2n}) = (-1)^P (1 \cdots 2n) \tag{1.24}$$

where P is the parity.

Examples

$$(1\ 2\ 3\ 4) = -(2\ 1\ 3\ 4) = - \begin{vmatrix} a_{21} & a_{23} & a_{24} \\ & a_{13} & a_{14} \\ & & a_{34} \end{vmatrix}$$

$$= - \begin{vmatrix} -(1\ 2) & (2\ 3) & (2\ 4) \\ & (1\ 3) & (1\ 4) \\ & & (3\ 4) \end{vmatrix} ;$$

$$(5\ 1\ 2\ 3\ 4\ 6) = + \begin{vmatrix} a_{51} & a_{52} & a_{53} & a_{54} & a_{56} \\ & a_{12} & a_{13} & a_{14} & a_{16} \\ & & a_{23} & a_{24} & a_{26} \\ & & & a_{34} & a_{36} \\ & & & & a_{46} \end{vmatrix}$$

$$= + \begin{vmatrix} -(1\ 5) & -(2\ 5) & -(3\ 5) & -(4\ 5) & (5\ 6) \\ & (1\ 2) & (1\ 3) & (1\ 4) & (1\ 6) \\ & & \ddots & & \vdots \\ & & & & (4\ 6) \end{vmatrix}$$

$$= + (1\ 2\ 3\ 4\ 5\ 6).$$

The *general rule* for the development of a pfaffian along line i is, in the fundamental ordering (1.23),

$$(1 \cdots 2n) = \sum_{i \neq j} (-1)^{1+i+j}(ij)(1, \ldots, i-1; i+1, \ldots, j-1;$$

$$j+1, \ldots, 2n) \tag{1.25}$$

if we take $(ij) = (ji)$. Or we can take the pfaffian with elements a_{hk}, permute i to be the first line according to (1.24), and expand with (1.5) to obtain, of course, the same result.

B. Number of Terms

The number of terms in the pfaffian $(1 \cdots 2n)$ is

$$(2n-1)!! = \frac{(2n)!}{2^n n!} = \frac{2^n \Gamma(n + \frac{1}{2})}{\Gamma(\frac{1}{2})}. \tag{1.26}$$

The meaning of the double factorial $(N!!)$ is $(2n)!! = 2^n n!$.

In this section we record, without proof, two relations between determinants and pfaffians. Any even antisymmetric determinant may be expressed in terms of pfaffians:

$$\begin{pmatrix} 1 \cdots 2n \\ 1 \cdots 2n \end{pmatrix} = (1 \cdots 2n)$$

where the elements of the pfaffian at the right-hand side are given by

$$(hk) = \begin{pmatrix} h & k \\ 1 & 2 \end{pmatrix} + \begin{pmatrix} h & k \\ 3 & 4 \end{pmatrix} + \cdots + \begin{pmatrix} h & k \\ 2n-1 & 2n \end{pmatrix}. \tag{1.27}$$

The following expansion holds.

$$(1 \cdots 2n) = \frac{(-1)^{\binom{n}{2}}}{2^n} \sum (-1)^P \begin{pmatrix} h_1 \cdots h_n \\ k_1 \cdots k_n \end{pmatrix}, \tag{1.28}$$

where in the indicated expansion P is the parity of the permutation $h_1 h_2 \cdots h_n, k_1 \cdots k_n$ of $1 \cdots 2n$, and $\sum$ is the sum over all $\binom{2n}{n}$ permutations such that

$$h_1 < h_2 < \cdots < h_n; \qquad k_1 < k_2 < \cdots < k_n.$$

Formula (1.28) also holds for *hafnians* (which are thereby expressed as sums of permanents) to be defined next.

7. HAFNIANS AND PERMANENTS

A hafnian is that counterpart of a pfaffian which obtains when all expansions carry the plus sign (instead of $\pm$),

$$[1 \cdots 2n] = \Sigma' [i_1 \, i_2][i_3 \, i_4] \cdots [i_{2n-1} \, i_{2n}], \qquad (1.29)$$

where $[i_1 \, i_2] = a_{i_1 \, i_2}$ and Σ' ranges over all permutations of $i_1 \cdots i_{2n}$ such that $i_1 < i_2, i_3 < i_4, \ldots$ and $i_1 < i_3 < \cdots < i_{2n-1}$. For example,

$$[1 \, 2 \, 3 \, 4] = [1 \, 2][3 \, 4] + [1 \, 3][2 \, 4] + [1 \, 4][2 \, 3].$$

Likewise, a *permanent* [5] is the counterpart of a determinant, when all terms in the expansion are taken with the plus sign. We shall write

$$\text{Perm } A = \begin{bmatrix} 1 \cdots m \\ 1 \cdots m \end{bmatrix} = \sum_{(i_1, \ldots, i_m)} [1 \, i_2] \cdots [m \, i_m]. \qquad (1.30)$$

We shall find it convenient always to denote hafnians and permanents, and their elements, with square brackets, and determinants and pfaffians with parentheses. Thus, square brackets will denote throughout "plus-sign" expansion rules; parentheses, "plus-or-minus-sign" expansion rules.

All expansion properties of hafnians and permanents that depend on combinatorics and term counting derive trivially from the corresponding ones that hold for pfaffians and determinants. Note, however, that the main connection between the latter—det $A = (\text{pf } A)^2$—*cannot* be generalized to hafnians and permanents.

Further algebraic work, which might enable us to connect the evaluation of large hafnians to something as simple as the evaluation of the eigenvalues of a determinant (this being the main use of the relation just quoted), would be of very great value; for instance, it would solve the N-dimensional Ising problem (of course, this also indicates the measure of its difficulty).

8. DE BRUIJN'S THEOREM

A remarkable connection between pfaffians and determinants is offered by a theorem due to N. G. de Bruijn [6], which deserves to be stated here, although it will not be needed in this context.

It is well known that

$$\int \cdots \int_{a \leqslant x_1 < \cdots < x_n \leqslant b} \det_{1 \leqslant i,j \leqslant n} [\phi_i(x_j)] \det_{1 \leqslant i,j \leqslant n} [\psi_i(x_j)] \, dx_1, \ldots, dx_n$$

$$= \det_{1 \leqslant i,j \leqslant n} \left[\int_a^b \phi_i(x)\, \psi_j(x) \, dx \right]; \tag{1.31}$$

de Bruijn's theorem states that, if in (1.31) we take only one (instead of two) integrand determinant factor, defining

$$\Omega = \int \cdots \int_{a \leqslant x_1 \cdots \leqslant x_2 \hat{m} \leqslant b} \det_{1 \leqslant i,j \leqslant 2m} [\phi_i(x_j)] \, dx_1, \ldots, dx_{2m} \tag{1.32}$$

(we have taken an even order in (1.32) for short), then

$$\Omega = \mathrm{pf}\, A \tag{1.33}$$

where A is the antisymmetric matrix of elements

$$a_{ij} = \int_a^b \int_a^b \phi_i(x)\phi_j(y)\, \mathrm{sgn}\,(y - x) \, dx \, dy.$$

Chapter 2

Miscellaneous Applications

In this chapter we shall consider some relevant applications of the pfaffian algorithm to various parts of science. This approach is instructive per se, and may stimulate further thought. The list is by no means exhaustive; the major omission is certainly that of the Ising model [7], which is probably also the most important use thus far made of pfaffians, outside the context of field theory: for this model the reader is referred to the works of E. W. Montroll [8] and P. W. Kasteleyn [9], and to the vast literature originated by them.

1. TRACES OF PRODUCTS OF DIRAC MATRICES [10]

A.

Let

$$P_i = \gamma^\mu P_\mu{}^{(i)} + 1 \cdot i p_5{}^{(i)} \tag{2.1}$$

be a vector matrix formed from the scalar $p_5{}^{(i)}$, the vector $p_\mu{}^{(i)}$, the unit matrix 1, and the Dirac matrices (in most cases $p_5{}^{(i)}$ is either a *mass term* or zero). Let $\mathcal{R} = P_1, P_2, \ldots, P_n$; we propose to evaluate the trace of $\mathcal{R}$, $\mathrm{Tr}(\mathcal{R})$. We shall first reduce $\mathcal{R}$ to a more convenient form by writing

$$\mathcal{R} = P_1 \gamma^5 \cdot \gamma^5 P_2 \cdot P_3 \gamma^5 \cdot \gamma^5 P_4 \cdots P_{2m-1} \gamma^5 \cdot \gamma^5 P_{2m}$$

$$\text{if} \quad n = 2m,$$

$$\mathcal{R} = P_2 \gamma^5 \cdot \gamma^5 P_2 \cdot P_3 \gamma^5 \cdot \gamma^5 P_4 \cdots P_{2m-1} \gamma^5 \cdot \gamma^5$$

$$\text{if} \quad n = 2m - 1. \tag{2.2}$$

and introducing the *dual* representation of the gamma matrices

$$\Gamma^5 = \gamma^5 = \gamma^1 \gamma^2 \gamma^3 \gamma^4; \qquad \Gamma^\mu = i\gamma^\mu \gamma^5; \qquad \Gamma^{\mu'} \Gamma^{\nu'} + \Gamma^{\nu'} \Gamma^{\mu'} = 2\delta^{\mu'\nu'}$$

$$(\mu = 1, \ldots, 4; \mu' = 1, \ldots, 5). \tag{2.3}$$

Any $\mathscr{R}$ may then be expressed as the product of an even number $2m$ of matrices

$$Q_h = \Gamma^{\mu'} q_{\mu'}{}^{(h)},$$

namely $(q_5{}^{(h)} = (-)p_5{}^{(h)}; p_1{}^{(h)}, p_2{}^{(h)}, p_3{}^{(h)}, p_4{}^{(h)} \equiv q_\mu{}^{(h)})$:

$$\mathscr{R} = Q_1 \cdots \cdots Q_{2m}.$$

Hence

$$\mathrm{Tr}(\mathscr{R}) = \tfrac{1}{4}\,\mathrm{Tr}(Q_1 \cdots \cdots Q_{2m}) = (1 \cdots 2m) \tag{2.4}$$

with

$$(hk) = \mathbf{q}^h \cdot \mathbf{q}^k = \sum_{\mu'=1}^{5} q_{\mu'}^{(h)} q_{\mu'}^{(k)}$$

Example. Consider the four matrices P_1, P_2, P_3, P_4 satisfying

$$P_r P_s + P_s P_r = 2(rs).$$

Since

$$\begin{aligned}
P_1 P_2 P_3 P_4 &= -P_2 P_1 P_3 P_4 + 2(1\,2)P_3 P_4 \\
&= P_2 P_3 P_1 P_4 + 2(\,1\,2)P_3 P_4 - 2(1\,3)P_2 P_4 \\
&= -P_2 P_3 P_4 P_1 + 2(1\,2)P_3 P_4 - 2(1\,3)P_2 P_4 + 2(1\,4)P_2 P_3
\end{aligned}$$

we obtain for the trace

$$\tfrac{1}{4}\,\mathrm{Tr}(P_1 P_2 P_3 P_4) = (1\,2)(3\,4) \quad (1\,3)(2\,4) + (1\,4)(2\,3).$$

B.

Consider the following product of h matrices Q_i:

$$\mathscr{R}_h = Q_1 \cdot Q_2 \cdots \cdots Q_h. \tag{2.5}$$

On introducing the 16 matrices of the Clifford algebra (these matrices were chosen to be Hermitian)

$$\Gamma_\epsilon^\alpha \equiv \{\Gamma_1{}^0 = 1;\ \Gamma_2{}^\mu = \gamma^\mu;\ \Gamma_3{}^{\mu\nu} = i\gamma^\mu \gamma^\nu;\ \Gamma_4{}^\mu = i\gamma^\mu \gamma^5;$$

$$\Gamma_5{}^0 = \gamma^5\}, \qquad \mu_1 \nu = 1, \ldots, q,$$

we clearly have

$$\mathscr{R}_h = \sum_{\alpha,\,\epsilon} (\mathscr{R}_h \Gamma_\epsilon{}^\mu) \Gamma_\epsilon{}^\alpha, \tag{2.6}$$

where each trace on the right-hand side of (2.6) is a pfaffian. (This
result is known as the reshuffling theorem.)

Furthermore,

$$\mathscr{R}_h^{\dagger} = Q_h Q_{h-1} \cdots Q_1 = \sum_{\alpha, \epsilon} (\mathscr{R}_h \Gamma_{\epsilon}^{\alpha})^{\dagger} \Gamma_{\epsilon}^{\alpha};$$

since $(\mathscr{R}_h \Gamma_{\epsilon}^{\alpha})$ is a number, it follows that

$$(\mathscr{R}_h \Gamma_{\epsilon}^{\alpha})^{\dagger} = (\mathscr{R}_h \Gamma_{\epsilon}^{\alpha})^*.$$

Of course, attention must be paid when relating such results to those
for $P_h P_{h-1} \cdots P_1$, because of the signs of $i p_5^{(j)}$ in the latter.

Expression (2.6) provides an alternative method for the *evaluation*
and *reduction* of traces. Writing

$$\mathscr{R}_h = \mathscr{R}_a \mathscr{R}_b \qquad (a + b = h),$$

we find

$$\mathscr{R}_h = \mathscr{R}_a \mathscr{R}_b = \sum_{\alpha, \epsilon} (\mathscr{R}_a \Gamma_{\epsilon}^{\alpha}) \Gamma_{\epsilon}^{\alpha} \sum_{\alpha', \epsilon'} (\mathscr{R}_b \Gamma_{\epsilon'}^{\alpha'}) \Gamma_{\epsilon'}^{\alpha'}$$

$$= \sum_{\alpha, \epsilon, \alpha', \epsilon'} (\mathscr{R}_a \Gamma_{\epsilon}^{\alpha})(\mathscr{R}_b \Gamma_{\epsilon'}^{\alpha'}) \Gamma_{\epsilon}^{\alpha} \Gamma_{\epsilon'}^{\alpha'};$$

but

$$\mathrm{Tr}(\Gamma_{\epsilon}^{\alpha} \Gamma_{\epsilon'}^{\alpha'}) = \delta_{\alpha \alpha'} \delta_{\epsilon \epsilon'},$$

so that

$$(\mathscr{R}_h) = \sum_{\alpha, \epsilon} (\mathscr{R}_a \Gamma_{\epsilon}^{\alpha})(\mathscr{R}_b \Gamma_{\epsilon}^{\alpha}) = \sum_{\alpha, \epsilon} (\mathscr{R}_a \Gamma_{\epsilon}^{\alpha})(\Gamma_{\epsilon}^{\alpha} \mathscr{R}_b). \tag{2.7}$$

Relevant simplifications occur when all $p_5^{(j)} \equiv 0$. To avoid confusion,
we shall call the matrices thus obtained

$$X_j = \sum_{\mu=1}^{4} \gamma^{\mu} p_{\mu}^{(j)}. \tag{2.8}$$

Consider now

$$Y_n = X_1 X_2 \cdots X_n$$

and distinguish between even and odd numbers of factors.

(*i*) *Even Case*

$$Y_{2m} = X_1 \cdots X_{2m} = (Y_{2m})1 + \sum_{\mu < \nu} (Y_{2m} i\gamma^{\mu}\gamma^{\nu}) i\gamma^{\mu}\gamma^{\nu}$$

$$+ (Y_{2m} \gamma^5)\gamma^5; \tag{2.9}$$

the vector and pseudo-vector contributions vanish, because traces of *odd* numbers of factors γ^m vanish. [Note: The last statement does not hold for the Γ^μ's. To see this, consider $(\Gamma^5\Gamma^1\Gamma^2\Gamma^3\Gamma^4) = (\Gamma^5\Gamma^5) = 1.$]

(ii) *Odd Case*

$$\overline{Y_{2m+1}} = X_1\cdots X_{2m+1}$$

$$= \sum_\mu (Y_{2m+1}\gamma^\mu)\gamma^\mu + \sum_\mu (Y_{2m+1}i\gamma^\mu\gamma^5)i\gamma^\mu\gamma^5. \tag{2.10}$$

Application of the formula

$$\sum_{p=1}^4 \gamma^p\mathscr{R}_h\gamma^p = 4(\mathscr{R}_h)1 - 2\sum_\mu (\mathscr{R}_h\gamma^\mu)\gamma^\mu$$

$$+ 2\sum_\mu (\mathscr{R}_h i\gamma^\mu\gamma^5)i\gamma^\mu\gamma^5 - 4(\mathscr{R}_h\gamma^5)\gamma^5 \tag{2.11}$$

to

$$\sum_{p=i}^4 \gamma^p Y_{2m}\gamma^p \quad\text{and}\quad \sum_p \gamma^p Y_{2m+1}\gamma^p \quad\text{yields}$$

$$\sum_{p=1}^1 \gamma^p Y_{2m}\gamma^p = 4(Y_{2m})1 - 4(Y_{2m}\gamma^5)\gamma^5,$$

$$\sum_{p=1}^4 \gamma^p Y_{2m+1}\gamma^p = -2\sum_\mu (Y_{2m+1}\gamma^\mu)\gamma^\mu + 2\sum_\mu (Y_{2m+1}i\gamma^\mu\gamma^5)i\gamma^\mu\gamma^5.$$

The latter expression can be simplified still further. Using

$$Y_n^\dagger = X_n\cdots X_2 X_1 = \sum_{\alpha_1\epsilon} (Y_n\Gamma_\epsilon{}^\alpha)^*\Gamma_\epsilon{}^\alpha, \tag{2.12}$$

we find, for $n = 2m+1$

$$Y_{2m+1}^\dagger = \sum_\mu (Y_{2m+1}\gamma^\mu)^*\gamma^\mu + \sum_\mu (Y_{2m+1}i\gamma^\mu\gamma^5)^*i\gamma^\mu\gamma^5,$$

which leads easily to

$$Y_{2m+1}^\dagger = \sum_\mu (Y_{2m+1}\gamma^\mu)\gamma^\mu - \sum_\mu (Y_{2m+1}i\gamma^\mu\gamma^5)i\gamma^\mu\gamma^5,$$

and finally to

$$\sum_P \gamma^p Y_{2m+1}\gamma^p = -2Y_{2m+1}^\dagger, \tag{2.13}$$

a well-known result.

In Chapter 3, Section 2, C we shall see that this formalism enables us to perceive on inspection that large numbers of terms vanish in a trace (2.4) when $n > 5$ if the vectors $\mathbf{P}^n$ are, as is usually the case, four-dimensional.

2. THE THEOREMS OF GAUSS, STOKES, AND LIOUVILLE

A. Oriented Volumes and Boundaries

The theorems of Gauss and Stokes, which play an important role in mathematical physics, may be expressed by means of Grassmann products in the concise form

$$\int_{\text{volume}} d\Omega = \int_{\text{boundary}} \Omega, \qquad (2.14)$$

where Ω is a differential form of any order; the differentials in (2.14) are not shown explicitly and are understood to be Grassmann products satisfying $dx^i \wedge dx^j = -dx^j \wedge dx^i$ with $dx^i \wedge dx^j = 0$ for $i, j = 1, 2, \ldots, n$. The sign of their products will be $+1$ or -1, depending on whether $1, 2, \ldots, n$ is an even or odd permutation: for example, $dx^1 \wedge dx^2 \wedge dx^3 = -dx^2 \wedge dx^1 \wedge dx^3$; $+1$ corresponds to *positive*, -1 to *negative* orientation of the volume element. We shall illustrate (2.14) by two examples.

Example 1. Suppose

$$\Omega = \omega_1 \, dx^2 \wedge dx^3 + \omega_2 \, dx^3 \wedge dx^1 + \omega_3 \, dx^1 \wedge dx^2;$$

then

$$d\Omega = \left(\frac{\partial \omega_1}{\partial x^1} + \frac{\partial \omega_2}{\partial x^2} + \frac{\partial \omega_3}{\partial x^3} \right) dx^1 \wedge dx^2 \wedge dx^3$$

$$= (\text{div } \omega) \, dx^1 \wedge dx^2 \wedge dx^3,$$

from which Gauss's theorem follows.

Example 2. Consider the quadratic form

$$\Omega = f_1 \, dx^1 + f_2 \, dx^2 + f_3 \, dx^3,$$

where f_1, f_2, and f_3 are functions of x_i, $i = 1, 2, 3$. Hence

$$d\Omega = \left(\frac{\partial f_2}{\partial x^1} - \frac{\partial f_1}{\partial x^2}\right) dx^1 \wedge dx^2 + \left(\frac{\partial f_1}{\partial x^3} - \frac{\partial f_3}{\partial x^1}\right) dx^3 \wedge dx^1$$

$$+ \left(\frac{\partial f_3}{\partial x^2} - \frac{\partial f_2}{\partial x^3}\right) dx^2 \wedge dx^3,$$

which leads to Stokes's theorem.

B. Liouville's Theorem

Consider a system with n degrees of freedom which is characterized by the canonical variables p_i, q_i with $i = 1, 2, \ldots, n$. Liouville's theorem asserts that the volume V of the $2n$-dimensional phase space,

$$V = \int \cdots \int dq_1 \cdots dq_n dp_1 \cdots dp_n \tag{2.15}$$

is an absolute invariant. In order to establish Liouville's theorem in terms of Grassmann products it suffices to consider the quadratic form

$$\Omega = \sum_{h=1}^{n} dq^h \wedge dp^h,$$

which is by definition invariant under canonical transformations (these form the *symmetry group* of ω). The nth power of ω, namely,

$$\Omega_{\wedge^n} = dq^1 \wedge \cdots \wedge dq^n \wedge dp^1 \wedge \cdots \wedge dp^n \tag{2.16}$$

is therefore likewise an invariant, so that Liouville's theorem follows immediately. The exponent in (2.16) is called the Grassmann *power*.

3. APPLICATIONS OF PFAFFIANS IN ECOLOGY [11]

A. The Volterra Equation

Suppose we have two competing species of fish, labeled prey (small fish) and predator (large fish). Let N_1 and N_2 denote the number of small and large fish, respectively. The differential equations satisfied by N_1 and N_2, in the classic Volterra model, are

$$\frac{dN_1}{dt} = \alpha_1 N_1 - \lambda_1 N_1 N_2, \qquad \frac{dN_2}{dt} = -\alpha_2 N_2 + \lambda_2 N_1 N_2,$$

where $-\lambda_1 N_1 N_2$ gives the rate of small fish being lost, while $+\lambda_2 N_1 N_2$ represents the growth rate of the population of larger fish. For n species the system of differential equations is of the form

$$\frac{dN_i}{dt} = k_i N_i + \frac{1}{\beta_i} \sum_{j=1}^{n} a_{ij} N_i N_j, \qquad i = 1, \ldots, n, \tag{2.17}$$

where $a_{ij} = -a_{ji}$ and $a_{ii} = 0$ for all i (no summation over i); the constants β_i are positive and are called equivalence numbers. Equations (2.17) are called the Volterra equations, and play an important role in ecology.

Suppose we are interested in the equilibrium solutions of system (2.17), in which case

$$\frac{dN_i}{dt} = 0 \qquad \text{for all } i.$$

If we further define $N_{i\,eq} \equiv q_i$, system (2.17) becomes

$$q_i \left(k_i \beta_i + \sum_{j=1}^{n} a_{ij} q_j \right) = 0,$$

or, provided $q_i \neq 0$ for $i = 1, \ldots, n$

$$\sum_{j=1}^{n} a_{ij} q_j = -k_i \beta_i, \qquad i = 1, \ldots, u. \tag{2.18}$$

If the number of species is odd, $n = 2N + 1$, no equilibrium solutions can exist, because $\det |a_{ij}| = 0$ (a_{ij} is an antisymmetric matrix of odd order). For example, if $i = 3$, we can easily check that $\det |a_{ij}| = 0$. Consequently *no* equilibrium solution exists among three species, if the initial population of each species is finite and different from zero.

B. Computation of Equilibrium Solutions

Simple and interesting algebraic considerations permit us to study and describe the situation when the *number of species is even*, $n = 2N$ [11]. Writing (2.18) in matrix notation, we have

$$A\mathbf{x} = \mathbf{b} \tag{2.19}$$

where A is an even antisymmetric matrix and $\mathbf{x}$ and $\mathbf{b}$ are $2N$-dimensional column matrices. Equation (2.19) is readily solved by Cramer's rule, provided of course that $\det A \neq 0$.

Thus

$$x_i \det A = \det B_i, \qquad i = 1, 2, \ldots, 2N, \tag{2.20}$$

$$x_i \det A = \begin{vmatrix} 0 & a_{12} & \cdots & a_{1,i-1} & b_{1i} & a_{1,i+1} & \cdots & a_{1,2N} \\ a_{21} & 0 & \cdots & a_{2,i-1} & b_{2i} & a_{2,a+1} & \cdots & a_{2,2N} \\ \vdots & & \ddots & & & & & \vdots \\ & & & \ddots & & & & \\ a_{i1} & & & & \ddots & & & \vdots \\ \vdots & & & & & \ddots & & \vdots \\ a_{2N,i} & \cdots & \cdots & & b_{2N,i} & \cdots & \cdots & 0 \end{vmatrix},$$

where $b_{ri} \equiv b_r$ (the index i has been inserted only as a reminder of x_i).
Let us compare now the two expansions by row i and column i:

$$\det A = \sum_{rs} a_{ir} a_{si} B_{ir}^{is} = \sum_{rs \neq i} (ir)(si) C_{ir}^{is},$$

$$a_{si} = \begin{cases} (si) & \text{if } s < i, \\ -(si) & \text{if } s > i, \end{cases} \tag{2.21}$$

and

$$\det B_i = \sum_{rs} a_{ir} b_{si} B_{ir}^{is} = \sum_{rs \neq i} (ir) b_{si}' C_{ir}^{is} \tag{2.22}$$

with

$$b_{si} \to b_{si} \equiv (si) = \begin{cases} b_{si} = b_s & \text{for } s < i, \\ -b_{si} & \text{for } s > i. \end{cases}$$

Furthermore

$$\det A = (\text{pf } A)^2 = \left[\sum_r (-1)^{1+i+r} (ir)(\text{pf } A)_{ir} \right]^2$$

$$= \sum_{r,s \neq i} (-1)^{1+i+s} (si)(\text{pf } A)_{si} (-1)^{1+i+r} (ir)(\text{pf } A)_{ir}$$

$$= \sum_{r,s \neq i} (-1)^{s+r} (si)(ir)(\text{pf } A)_{si} (\text{pf } A)_{ir}. \tag{2.23}$$

It easily follows that

$$x_i (\text{pf } A) = (1, \ldots, i-1; b_i'; i+1, \ldots, 2N) \tag{2.24}$$

Formula (2.24) shows explicitly that Cramer's rule reduces in this case to the ratio of two pfaffians. It might also serve as a noteworthy definition of pfaffian.

Example   Consider a system of four simultaneous equations in the unknowns x_i, $i = 1, \ldots, 4$, so that $N = 2$ here. Then

$$x_1(1\ 2\ 3\ 4) = (b'\ 2\ 3\ 4) = \begin{vmatrix} -b_2 & -b_3 & -b_4 \\ & (2\ 3) & (2\ 4) \\ & & (3\ 4) \end{vmatrix},$$

$$x_2(1\ 2\ 3\ 4) = (1\ b'\ 3\ 4) = \begin{vmatrix} b_1 & (1\ 3) & (1\ 4) \\ & -b_3 & -b_4 \\ & & (3\ 4) \end{vmatrix},$$

$$x_3(1\ 2\ 3\ 4) = (1\ 2\ b'\ 4) = \begin{vmatrix} (1\ 2) & b_1 & (1\ 4) \\ & b_2 & (2\ 4) \\ & & -b_4 \end{vmatrix},$$

$$x_4(1\ 2\ 3\ 4) = (1\ 2\ 3\ b') = \begin{vmatrix} (1\ 2) & (1\ 3) & b_1 \\ & (2\ 3) & b_2 \\ & & b_3 \end{vmatrix}.$$

This enables us to see that only the horizontal part of the line b' changes sign with respect to the values b; in general,

$$\begin{array}{c|l} + & \leftarrow \textit{Vertical} \text{ part of line } i \\ + & \\ + & \textit{Horizontal} \text{ part of line } i \\ + & \downarrow \\ \hline \end{array}$$

— — — — — — — — —

4. FORMAL SOLUTIONS OF FREDHOLM'S INTEGRAL EQUATION

A. General Considerations

The solution of the integral equation

$$\phi(x) = f(x) + \lambda \int_0^1 K(x, y)\phi(y)\, dy \tag{2.25}$$

was first given by Fredholm in the form

$$\phi(x) = f(x) + \lambda \int_0^1 \Gamma(x, y \mid \lambda) f(y) \, dy, \tag{2.26}$$

where the *resolvent* $\Gamma(x, y \mid \lambda)$ is

$$\Gamma(x, y \mid \lambda) = \frac{D(x, y \mid \lambda)}{D_0(\lambda)}. \tag{2.27}$$

$K(x, y)$ is called the *kernel*. The functions $D(x, y \mid \lambda)$ and $D_0(\lambda)$ possess the following series representations.

$$D(x, y \mid \lambda) = \sum_{m=0}^{\infty} \frac{(-\lambda)^n}{n!} \int_0^1 d\xi_1 \cdots \int_0^1 d\xi_u \begin{pmatrix} x & \xi_1 & \cdots & \xi_u \\ y & \xi_1 & \cdots & \xi_u \end{pmatrix} \tag{2.28a}$$

and

$$D_0(\lambda) = \sum_{n=0}^{\infty} \frac{(-\lambda)^n}{n!} \int_0^1 d\xi_1 \cdots \int_0^1 d\xi_n \begin{pmatrix} \xi_1 & \cdots & \xi_n \\ \xi_1 & \cdots & \xi_n \end{pmatrix}, \tag{2.28b}$$

where the elements of the determinant

$$\begin{pmatrix} x_1 & \cdots & x_n \\ y_1 & \cdots & y_n \end{pmatrix}$$

are

$$(x_i y_j) = K(x_i, y_j). \tag{2.29}$$

It is interesting to note that $\Gamma(x, y \mid \lambda)$ may also be written in terms of permanents [12] and of hafnians [13] if the kernal is symmetric. To this effect consider

$$D(\lambda) = \exp[- \int_0^\lambda \delta(\lambda) \, d\lambda] \tag{2.30}$$

with

$$\delta(\lambda) = \int_0^1 \Gamma(x, x \mid \lambda) \, dx;$$

$\Gamma(x, y \mid \lambda)$ may then be expressed as the ratio

$$\Gamma(x, y \mid \lambda) = \frac{P(x, y \mid \lambda)}{P(\lambda)}; \tag{2.31}$$

here

$$P(x, y \mid \lambda) = \sum_{n=0}^{\infty} \frac{\lambda^n}{n!} \int_0^1 d\xi_1 \cdots \int_0^1 d\xi_n \begin{bmatrix} x & \xi_1 & \cdots & \xi_n \\ y & \xi_1 & \cdots & \xi_n \end{bmatrix}, \qquad (2.32a)$$

$$P(\lambda) = \sum_{u=0}^{\infty} \frac{\lambda^n}{u!} \int_0^1 d\xi_1 \cdots \int_0^1 d\xi_u \begin{bmatrix} \xi_1 & \cdots & \xi_u \\ \xi_1 & \cdots & \xi_u \end{bmatrix}, \qquad (2.32b)$$

and $[hk] = K(\xi_h, \xi_k)$. The following relations hold between the functions P and D.

$$P(\lambda)D(\lambda) = 1, \qquad (2.33a)$$

$$P(x, y \mid \lambda) = [P(\lambda)]^2 D(x, y \mid \lambda). \qquad (2.33b)$$

B. Symmetrical Kernels

For a symmetrical kernel

$$K(x, y) = K(y, x) \qquad (2.34)$$

the resolvent $\Gamma(x, y \mid \lambda)$ may be written as a ratio of two series of hafnians:

$$\Gamma(x, y \mid \lambda) = \frac{H(x, y \mid \lambda)}{H(\lambda)}, \qquad (2.35)$$

where

$$H(x, y \mid \lambda) = \sum_{n=0}^{\infty} \left(\frac{\lambda}{2}\right)^n \frac{1}{n!} \int_0^1 d\xi_1 \cdots \int_0^1 d\xi_n \, [xy \, \xi_1 \xi_1 \xi_2 \xi_2 \cdots \\ \xi_n \xi_u], \qquad (2.36a)$$

$$H(\lambda) = \sum_{n=0}^{\infty} \left(\frac{\lambda}{2}\right)^n \frac{1}{n!} \int_0^1 d\xi_1 \cdots \int_0^1 d\xi_n \, [\xi_1 \xi_1 \cdots \xi_u \xi_u], \qquad (2.36b)$$

and $[hk] = [kh] = K(\xi_h, \xi_k) = K(\xi_k, \xi_h)$.

An interesting result is the following relation between $P(\lambda)$ and $H(\lambda)$, these quantities being given respectively by Eqs. (2.32b) and (2.36b).

$$H(\lambda) = (+P(\lambda))^{\frac{1}{2}}, \qquad (2.37)$$

Chapter 3

Expansion Theorems

1. TRANSITION FROM GRASSMANN ALGEBRAS TO CLIFFORD ALGEBRAS (A DIGRESSION)

In Chapter 1 the Grassmann and Clifford products were defined respectively by

$$x^h \wedge x^k = -x^k \wedge x^h, \qquad x^h \wedge x^h = 0, \tag{3.1a}$$

and

$$x^h \wedge x^k = -x^k \wedge x^h + 2\delta_{hk}, \qquad x^h \wedge x^h = 1. \tag{3.1b}$$

We shall show now that, given a Grassmann algebra, it is always possible to find a corresponding Clifford algebra. The proof of this theorem is based on defining the following differentiation law for Grassmann products.

$$\partial_h \wedge x^k = -x^k \wedge \partial_h, \qquad h \neq k,$$

$$\partial_h \wedge \partial_k = -\partial_k \wedge \partial_h. \tag{3.2}$$

Equation (3.2) takes, when the case $h = k$ is included, the suggestive form

$$\partial_h \wedge x^k = \delta_{hk} - x^k \wedge \partial_h, \tag{3.3}$$

δ_{hk} being the Kronecker symbol. If we introduce the new variables y_h

$$y_h = x_h + \partial_h, \tag{3.4}$$

it immediately follows that

$$y_h \wedge y_k = -y_k \wedge y_h + 2\delta_{hk}. \tag{3.5}$$

Comparison of (3.5) with (3.1b) indicates clearly that the elements y_h

obey the multiplication law of the Clifford algebra (correctly speaking, the symbol $\wedge$ in (3.5) should now be replaced by $\underline{\wedge}$).

We shall not pursue further here these considerations which would easily lead to results of more than formal interest; their extension to commuting variables (Bose fields, see Section 3) is immediate.

2. RULES CONCERNING DETERMINANTS, PFAFFIANS, PERMANENTS, AND HAFNIANS

A. Determinants

We shall present a little-known expansion rule for determinants, due to Arnaldi, [2] which has proved essential especially in our combinatoric treatment of renormalization theory. To start with an example, consider the nth-order determinant

$$D = \begin{vmatrix} a & b & c & d & \cdots \\ a' & b' & c' & d' & \cdots \\ a'' & b'' & c'' & d'' & \cdots \\ \cdot & \cdot & \cdot & \cdot & \cdots \end{vmatrix}$$

and its minor

$$M = \begin{vmatrix} a & b \\ a' & b' \end{vmatrix}.$$

Arnaldi's idea was to expand D explicitly "in terms of the elements contained in the minor M." To see this, note first of all that D may always be expressed as the sum of two determinants, provided we confine ourselves to either a row or a column. If we "work with column 1," D becomes

$$D = \begin{vmatrix} 0 & b & \cdots \\ 0 & b' & \cdots \\ a'' & b'' & \cdots \\ a''' & b''' & \cdots \\ \cdot & \cdot & \cdots \end{vmatrix} + \begin{vmatrix} a & b & \cdots \\ a' & b' & \cdots \\ 0 & b'' & \cdots \\ 0 & b''' & \cdots \\ \cdot & \cdot & \cdots \end{vmatrix}.$$

This technique of introducing zeros by writing D as a sum of determinants may be continued until

$$
D = \begin{vmatrix} 0 & 0 & \vdots & \cdots \\ 0 & 0 & \vdots & \cdots \\ \hline \cdot & \cdot & & \cdots \\ \cdot & \cdot & \cdot & \cdots \end{vmatrix} + \Sigma \pm \alpha \underbrace{\begin{vmatrix} 0 & \cdots \\ \vdots & \ddots \\ \vdots & \vdots & \ddots \end{vmatrix}}_{\text{order } n-1} + \overbrace{\begin{vmatrix} a & b \\ a' & b' \end{vmatrix} \begin{vmatrix} \cdots \\ \cdots \\ \cdots \end{vmatrix}}^{\text{order } n-2}. \tag{3.6}
$$

$$(\alpha = a, a', b, b')$$

Since the original Arnaldi expansion (3.6) is rather clumsy, we introduce a notation to "streamline" it. To this effect let us use the notation $(hk) = a_{hk}$ and define the following symbols.

$$(\overset{\circ}{h}\overset{\circ}{k}) = 0 \tag{3.7a}$$

and

$$(\overset{\circ}{h}k) = (h\overset{\circ}{k}) = (hk). \tag{3.7b}$$

Example 1.   Consider the fourth-order determinant

$$
\begin{pmatrix} 1\ 2\ 3\ 4 \\ 1\ 2\ 3\ 4 \end{pmatrix} = \begin{vmatrix} (1\ 1) & (1\ 2) & (1\ 3) & (1\ 4) \\ (2\ 1) & \cdot & \cdot & \cdot \\ \vdots & & & \vdots \\ (4\ 1) & \cdot & \cdot & (4\ 4) \end{vmatrix} ;
$$

the meaning of

$$
\begin{pmatrix} \overset{\circ}{1}\ \overset{\circ}{2}\ 3\ 4 \\ \overset{\circ}{1}\ \overset{\circ}{2}\ 3\ 4 \end{pmatrix}
$$

then is

$$
\begin{pmatrix} \overset{\circ}{1}\ \overset{\circ}{2}\ 3\ 4 \\ \overset{\circ}{1}\ \overset{\circ}{2}\ 3\ 4 \end{pmatrix} = \begin{vmatrix} (\overset{\circ}{1}\ \overset{\circ}{1}) & (\overset{\circ}{1}\ \overset{\circ}{2}) & (1\ 3) & (1\ 4) \\ (\overset{\circ}{2}\ \overset{\circ}{1}) & (\overset{\circ}{2}\ \overset{\circ}{2}) & (2\ 3) & (2\ 4) \\ (3\ 1) & (3\ 2) & (3\ 3) & (3\ 4) \\ (4\ 1) & (4\ 2) & (4\ 3) & (4\ 4) \end{vmatrix}
$$

$$
= \begin{vmatrix} 0 & 0 & (1\ 3) & (1\ 4) \\ 0 & 0 & (2\ 3) & (2\ 4) \\ (3\ 1) & (3\ 2) & (3\ 3) & (3\ 4) \\ (4\ 1) & (4\ 2) & (4\ 3) & (4\ 4) \end{vmatrix}.
$$

We are now ready to express Arnaldi's expansion rule in a conveniently compact form.

Theorem. Consider the first m rows and columns of the nth-order determinant

$$\begin{pmatrix} 1, 2, \ldots, n \\ 1, 2, \ldots, n \end{pmatrix};$$

then

$$\begin{pmatrix} 1, \ldots, m; m+1, \ldots, n \\ 1, \ldots, m; m+1, \ldots, n \end{pmatrix} = \sum_{r=0}^{m} \sum_{C_r^{\text{row}}} \sum_{C_r^{\text{col}}} (-1)^{P(h,k)}$$

$$\times \begin{pmatrix} h'_1, \ldots, h'_r \\ k'_1, \ldots, k'_r \end{pmatrix} \begin{pmatrix} \overset{\circ}{h''_1}, \ldots, \overset{\circ}{h''_s}; m+1, \ldots, n \\ \overset{\circ}{k''_1}, \ldots, \overset{\circ}{k''_s}; m+1, \ldots, n \end{pmatrix}, \tag{3.8}$$

where

$$h'_1 < h'_2 < \cdots < h'_r; \qquad h''_1 < \cdots < h''_s;$$

$$k'_1 < \cdots < k'_r, \qquad k''_1 < \cdots < k''_s; \qquad r+s = m;$$

C_r^{row} and C_r^{col} denote, respectively, all possible row and column permutations; and $P(h, k)$ is defined as the parity of the row indices with respect to the column indices.

B. Permanents

These objects, which we denoted by square brackets, are similar to determinants, except that each term in the development of a permanent is written with a *plus* sign. For example

$$\begin{bmatrix} 1 & 2 & 3 & 4 \\ 1 & 2 & 3 & 4 \end{bmatrix}$$

is a permanent of order 4 with elements [1 2], [2 4], and so on. It follows from these considerations that Arnaldi's expansion rule is also applicable to permanents, provided we write each term in the expansion with a plus sign and convert all determinants into permanents.

C. Pfaffians

In order to obtain a similar expansion for pfaffians, we single out a set L of m lines, which we label with indices h (a line consists in general of a horizontal and vertical part). The expansion then has the form [15]

$$(1 \cdots 2n) = \sum_{r=0}^{[m/2]} (-1)^{\binom{m-2r}{r}} \sum_{C_r} (-1)^{P'(h,\,k)} (h'_1, \ldots, h'_{2r})$$

$$\times \begin{pmatrix} h''_1, \ldots, h''_s \\ k''_1, \ldots, k''_s \end{pmatrix} (k'_1, \ldots, k'_{2t}), \tag{3.9}$$

where $P'(h, k)$ is the parity of $h'_1, \ldots, h'_{2r}, h''_1, \ldots, h''_s, k''_1, \ldots, k''_s,$ $k'_1, \ldots, k'_{2t}$ with respect to $1, 2, \ldots, 2n$; C_r denotes all possible combinations of $2r$ out of the m fixed lines; $m = 2r + s$.

Next we remove $2r$ h's from L' and write them in natural order $(h'_1, \ldots, h'_{2r})$. Finally, since we want $2r + s = m$, we take s columns from the lines L and write the k's in all possible ways such that

$$s + 2t = 2n - m.$$

With the notation (3.7), the expansion (3.9) reads

$$(1 \cdots 2n) = \sum_{r=0}^{[m/2]} \sum_{C_r} (-1)^{P(h)} (h'_1 \ldots h'_{2r})$$

$$\times (\overset{\circ}{h}''_1 \ldots \overset{\circ}{h}''_s; m + 1, \ldots, 2n). \tag{3.10}$$

An important application of (3.9) arises when all pfaffians on its right-hand side vanish and only a determinant of order n is left (e.g., in quantum electrodynamics). It is instructive to apply this formula, as an example, to the computation of traces; remarkable simplifications are exhibited that would remain hidden otherwise.

Example 2.   We recall our rule for traces (1.26) where

$$(1 \cdots 2n) = \text{Tr}(Q_1 Q_2 \cdots Q_{2n})$$

with

$$(hk) \equiv \mathbf{q}^h \cdot \mathbf{q}^k.$$

Taking the expansion (3.9) for the first m lines, we have

$$2r + s = m \qquad \text{or} \qquad s = m - 2r.$$

As soon as $s > 5$, *all determinants of the type*

$$D = \begin{pmatrix} h_1'' \cdots h_s'' \\ k_1'' \cdots k_s'' \end{pmatrix}$$

vanish. To see this we note that since $(hk) = \mathbf{q}^h \cdot \mathbf{q}^k$ $(h, k = 1, \ldots, s)$, D may be written as a matrix product

$$\begin{pmatrix} 1 \cdots s \\ 1 \cdots s \end{pmatrix} = \begin{pmatrix} q_1^{(1)} & \cdots & q_s^{(1)} \\ \vdots & & \vdots \\ q_1^{(s)} & \cdots & q_s^{(s)} \end{pmatrix} \begin{pmatrix} q_1^{(1)} & \cdots & q_1^{(s)} \\ \vdots & & \vdots \\ q_s^{(1)} & \cdots & q_s^{(s)} \end{pmatrix}$$

and therefore its rank cannot exceed 5.

D. Hafnians

The same remarks apply for hafnians as in the expansion of permanents: formula (3.9) shall be taken when all signs are plus signs, and square brackets throughout.

Formula (3.10) thus becomes, with the same combinatoric notation,

$$[1 \cdots 2n] = \sum_{n=0}^{\lfloor m/2 \rfloor} \sum_{C_r} [h_1', \ldots, h_{2r}'] [\overset{\circ}{h_1''}, \ldots, \overset{\circ}{h_s''}, m+1, \ldots, 2n]. \tag{3.11}$$

We also record here a remarkable expansion (among the many that can be proved) [1, 15, 16] because of its interest in our later discussion of renormalization (cf. Chapter 9). Let $[hk] = [kh]$; denote with $\sum_{C_\rho}$ the sum over all the C_ρ combinations $h_1 < h_2 < \cdots < h_{2\rho}$, $k_1 < k_2 < \cdots < k_\sigma$ of the indices $1, 2, \ldots, \mu$ $(\mu = 2\rho + \sigma)$, and with $\sum_{C_r}$ the sum over all C_r combinations $l_1 < \cdots < l_\sigma$, $m_1 < \cdots < m_{2r}$ of the indices $\mu + 1$, $\mu + 2, \ldots, 2n$ $(2n - \mu = 2r + \sigma)$; then

$$\sum_{C_\rho} [h_1, \ldots, h_{2\rho}] [\overset{\circ}{k_1}, \ldots, \overset{\circ}{k_\sigma}; \mu + 1, \ldots, 2n]$$

$$= \sum_{C_r} [1, \ldots, \mu; \overset{\circ}{l_1}, \ldots, \overset{\circ}{l_\sigma}] [m_1, \ldots, m_{2r}]. \tag{3.12}$$

When $\sigma = 2$, this becomes

$$\sum_{C_r} [l_1 l_2] [m_1 \cdots m_{2r}] = n[1 \cdots 2n]. \tag{3.12'}$$

3. BASIC COMBINATORICS FOR PERTURBATIVE EXPANSIONS

We recall from Chapter 1 the fundamental relation between Clifford's algebra (where multiplication is denoted by $\wedge$) and Grassmann's algebra:

$$\omega_1 \wedge \cdots \wedge \omega_k = \sum_{r=0}^{[k/2]} \sum_{C_r} (-1)^P (h_1 \cdots h_{2r}) \, \omega_{l_1} \wedge \omega_{l_2}$$

$$\mathbf{x} \cdots \wedge \omega_{l_{k-2r}}, \tag{3.13}$$

where $[k/2]$ is the greatest integer function and $(h_1 \cdots h_{2r})$ is a pfaffian with $(h_1 h_2) = (h_2 h_1) = \Sigma_i a_{h_1 i} a_{h_2 i}$. For the simple case where $k = 2$, Eq. (3.13) becomes

$$\omega_1 \wedge \omega_2 = (1\ 2) + \omega_1 \wedge \omega_2. \tag{3.13'}$$

We note, for later use, that this expression is of the same form as *Wick's theorem* for field operators A, B [17-21]:

$$T(AB) = \overline{AB} + N(AB), \tag{3.14}$$

where $\overline{AB}$ are the contracted factors; T and N denote, respectively, the time-ordered and normal products of A, B.

We also make the *fundamental remark* that the demonstration of (3.13) (omitted in this book) depends solely upon the validity of (3.13') with $\omega_1 \wedge \omega_2 = -\omega_2 \wedge \omega_1$. Hence we shall be able to use (3.13) *directly*, as soon as a decomposition like (3.13') or (3.14) is given for only two fields.

Clifford products of the type (3.13), (3.13') occur when we deal with creation and annihilation operators of a fermion field. The question arises, do expressions similar to (3.13) also exist for boson fields? The answer is yes, provided we define the following multiplication law on the set of elements $x_1, \ldots, x_n$.

$$x_h \vee x_k = \begin{cases} x_k \vee x_h & \text{for } h \neq k, \\ 0 & \text{for } h = k, \end{cases} \tag{3.15}$$

and

$$x_h \mathbin{\mathpalette\@vee\relax} x_k = \begin{cases} x_k \mathbin{\mathpalette\@vee\relax} x_h & \text{for } h \neq k, \\ 1 & \text{for } h = k. \end{cases} \tag{3.16}$$

The products (3.15) and (3.16) are the analogues of the Grassmann

($\wedge$) and Clifford ($\overline{\wedge}$) products. We are now in a position to give the corresponding expression for *boson fields:*

$$\omega_1 \vee \cdots \vee \omega_k = \sum_{r=0}^{[k/2]} \sum_{C_r} [h_1 \cdots h_{2r}] \omega_{l_1} \vee \omega_{l_2} \cdots \vee \omega_{l_k - 2r}. \quad (3.17)$$

Equation (3.17) differs formally from Eq. (3.13) only in the replacement of

(i)   pfaffians $(1 \cdots 2n) \rightarrow$ hafnians $[1 \cdots 2n]$;
(ii)   the Grassmann symbol $\wedge \rightarrow \vee$;
(iii)   the Clifford symbol $\overline{\wedge} \rightarrow \overline{\vee}$.

We conclude that determinants and pfaffians are suitable tools for fermion fields, whereas permanents and hafnians are to be used when dealing with boson fields. We note that the rule (3.16) is introduced only as a formal trick to utilize without changes the result already obtained for fermions; all the developments used later could also be obtained replacing (3.16) with $x_h \vee x_k = x_k \vee x_h$ for all h, k.

4. THE PHOTON FIELD (VECTOR FIELD)

Consider two boson fields $A_{\mu_1}(x_1)$ and $A_{\mu_2}(x_2)$; we can then take as "Clifford product" the ordinary product and get

$$A_{\mu_1}(x_1) A_{\mu_2}(x_2) = [A_{\mu_1}^+(x_1) + A_{\mu_1}^-(x_1)][A_{\mu_2}^+(x_2) + A_{\mu_2}^-(x_2)]$$

$$= A_{\mu_1} \vee A_{\mu_2} + [1\ 2]_+ \quad (3.18)$$

where

$$A_{\mu_1}(x_1) \vee A_{\mu_2}(x_2) \equiv\ : A_{\mu_1}(x_1) A_{\mu_2}(x_2): \quad (3.19a)$$

denotes *normal* ordering in the sense of Wick [17–21], while

$$[1\ 2]_+ = [A_{\mu_1}^+(x_1), A_{\mu_2}^-(x_2)]_- \quad (3.19b)$$

is the Wick contraction. Taking the vacuum expectation value of (3.18), we obtain

$$\langle 0 | A_{\mu_1}(x_1) A_{\mu_2}(x_2) | 0 \rangle = \langle 0 | A_{\mu_1} \vee A_{\mu_2} | 0 \rangle + \langle 0 | [1\ 2]_+ | 0 \rangle$$

$$= [1\ 2]_+ = i D_{\mu_1 \mu_2}^+(x_1 - x_2), \quad (3.20)$$

since

$$\langle 0 | A_{\mu_1} \vee A_{\mu_2} | 0 \rangle = 0.$$

For k boson fields, as was said in the previous section, we can use (3.17) directly:

$$A_{\mu_1}(x_1) \cdots A_{\mu_k}(x_k) = \sum_{r=0}^{[k/2]} \sum_{C_r} [h_1 \cdots h_{2r}]_+$$

$$\times : A_{\mu_{l_1}}(x_1) \cdots A_{\mu_{l_{k-2r}}}(x_{l_{k-2r}}): \qquad (3.21)$$

Formulas (3.20) and (3.21) are easily generalized to T products, defined by

$$T(A_{\mu_1}(x_1) A_{\mu_2}(x_2)) = \begin{cases} A_1 A_2 & \text{for } x_1^0 > x_2^0, \\ A_2 A_1 & \text{for } x_2^0 > x_1^0; \end{cases}$$

that is,

$$T(A_{\mu_1}(x_1) A_{\mu_2}(x_2)) = \theta(1\text{-}2) A_{\mu_1}(x_1) A_{\mu_2}(x_2)$$
$$+ \theta(2\text{-}1) A_{\mu_2}(x_2) A_{\mu_1}(x_1), \qquad (3.22)$$

with the standard step function

$$\theta(x) = \begin{cases} 1 & \text{if } x^0 > 0, \\ 0 & \text{if } x^0 < 0. \end{cases}$$

The time-ordered product (3.22) can be written concisely as

$$T(A_{\mu_1}(x_1) A_{\mu_2}(x_2)) = [1\ 2] + : A_{\mu_1}(x_1) A_{\mu_2}(x_2): \qquad (3.23)$$

where

$$[1\ 2] = \theta(x_1 - x_2)[1\ 2]_+ + \theta(x_2 - x_1)[2\ 1]_+$$
$$= \tfrac{1}{2} D^F_{\mu_1 \mu_2}(x_1 - x_2) \qquad (3.24)$$

is known as the Feynman causal propagator. The vacuum expectation value of (3.23) is

$$\langle 0 | T(A_{\mu_1}(x_1) A_{\mu_2}(x_2)) | 0 \rangle = [1\ 2]. \qquad (3.25)$$

We have then (cf. (3.21): as Clifford product we take now the T product)

$$T(A_{\mu_1}(x_1) \cdots A_{\mu k}(x_k)) = \sum_{r=0}^{[k/2]} \sum_{C_r} [h_1 \cdots h_{2r}] : A_{\mu_{l_1}}(x_1) \cdots$$

$$\times A_{\mu_{l_{k-2r}}}(x_{l_{k-2r}}): \qquad (3.26)$$

which is the central result for our purposes.

It is furthermore easy to handle T products containing partial Wick orderings, such as

$$T(:1\ 2\ 3:\ :4\ 5:\ \cdots :6\ 7\ 8\ 9:)$$

where $k \equiv A_{\mu_k}(x_k)$: it suffices to suppress all terms in the expansion of the hafnians in (3.26) that link points contained within the same Wick product, such as $[1\ 2]$, $[6\ 8]$, and so on.

Entirely analogous results occur for ordinary and T products of fermion field operators, in terms of pfaffians and determinants. They will be omitted for brevity's sake, except for those needed in the perturbative expansions which are reported in the next section.

We wish to record, in conclusion, two formulas that invert, respectively, (3.21) and (3.26):

$$:A_{\mu_1}(x_1) \cdots A_{\mu_n}(x_n): = \sum_{r=0}^{[n/2]} (-1)^r \sum_{C_r} [h_1 \cdots h_{2r}] A_{\mu_{k_1}}(x_{k_1})$$

$$\cdots A_{\mu_{k_s}}(x_{k_s}) \tag{3.27}$$

and

$$:A_{\mu_1}(x_1) \cdots A_{\mu_n}(x_n): = \sum_{r=0}^{[n/2]} (-1)^r [h_1 \cdots h_{2r}]$$

$$\times T(A_{\mu_{k_1}}(x_{k_1}) \cdots A_{\mu_{k_s}}(x_{k_s})). \tag{3.28}$$

where $h_1 < \cdots < h_{2r}$, and $k_1 < \cdots < k_s$ denote the remaining variables.

5. PERTURBATIVE EXPANSIONS IN QUANTUM ELECTRODYNAMICS

Our aim is to express the nth-order term in the S-matrix expansion in concise form. If the interaction Lagrangian in quantum electrodynamics is written as

$$i\lambda \bar{\psi}(x) \gamma^\mu \psi(x) A_\mu(x),$$

the expansion of Dyson's U matrix [17–21] reads

$$U(t_f, t_i) = \sum_{N=0}^{\infty} \frac{\lambda^N}{N!} \int_{t_i}^{t_f} d^4x_1 \int_{t_i}^{t_f} d^4x_2 \cdots \int_{t_i}^{t_f} d^4x_N$$

$$T\left[\prod_{i=1}^{N} \bar{\psi}_{\alpha_i}(x_i) \psi_{\beta_i}(x_i) A_{\mu_i}(x_i) \gamma^{\mu_i}_{\alpha_i \beta_i}\right], \tag{3.29}$$

where λ is the coupling constant, while t_f and t_i denote, respectively, final and initial times; $\bar{\psi}(x) = \psi^*(x)\gamma^4$.

Let us start with the vacuum–vacuum transition amplitude

$$M_{00} \equiv \langle 0 \,|\, U \,|\, 0 \rangle;$$

it is necessary to evaluate first $\langle 0 \,|\, T$ (boson fields) $| 0 \rangle$ and $\langle 0 \,|\, T$ (fermion fields) $| 0 \rangle$.

(i) In the case of bosons, only $N = 2M$ gives a result not equal to 0. This follows directly from Eq. (3.26), which is seen to yield a nonzero vacuum expectation value only if *all* A_μ's are contracted. Recalling (3.26), we obtain the *hafnian*

$$\langle 0 \,|\, T(A_{\mu_1}(x_1) \cdots A_{\mu_{2n}}(x_{2n})) | 0 \rangle = [1 \cdots 2n]. \tag{3.30}$$

(ii) For a product of *fermion* field operators such as

$$\bar{\psi}_{\alpha_1}(x_1)\psi_{\beta_1}(x_1) \cdots \bar{\psi}_{\alpha_N}(x_N)\psi_{\beta_N}(x_N),$$

the vacuum expectation value of the corresponding T product is given by the pfaffian

$$\langle 0 \,|\, T(\bar{\psi}_{\alpha_1}(x_1)\psi_{\beta_1}(x_1) \cdots \bar{\psi}_{\alpha_N}(x_N)\psi_{\beta_N}(x_N)) | 0 \rangle$$
$$= (\bar{1}\, 1\, \bar{2}\, 2 \cdots \bar{N}\, N), \tag{3.31}$$

which has all the elements

$$(\bar{r}\bar{s}) = 0 \rightarrow \overline{\bar{\psi}_{\alpha_r}(x_r)\bar{\psi}_{\alpha_s}(x_s)} = 0,$$

$$(r's') = 0 \rightarrow \overline{\psi_{\beta_{r'}}(x_{r'})\psi_{\beta_{s'}}(x_{s'})} = 0$$

and therefore reduces [1] by virtue of our expansion (3.9) to a *determinant*

$$(\bar{1}\, 1 \cdots \bar{N}\, N) = \begin{pmatrix} 1 \cdots N \\ 1 \cdots N \end{pmatrix}, \tag{3.32}$$

which can be shown to have elements

$$(hk) = \tfrac{1}{2}S^F_{\beta_h \alpha_k}(x_h - x_k), \tag{3.33}$$

S^F being the Feynman propagator for the free fermion field.

Returning to the amplitude M_{00}, we find, using Eqs. (3.30) and (3.31), (3.32) that

$$M_{00} = \langle 0 | v(t_f, t_i) | 0 \rangle$$

$$= \sum_{n=0}^{\infty} \frac{\lambda^{2n}}{(2n)!} \int d^4 x_1 \cdots \int d^4 x_{2n} \gamma^1 \cdots \gamma^{2n} \begin{pmatrix} 1 \cdots 2n \\ 1 \cdots 2n \end{pmatrix}$$

$$\times [1 \cdots 2n],$$

$$\gamma^1 = \gamma^{\mu_1}_{\alpha_1 \beta_1}, \text{ etc.,} \tag{3.34}$$

which gives the contributions of all Feynman diagrams without external lines.

In all generality, it can be shown [1] that the expectation value of any product of Bose (Fermi) fields between *arbitrary* states is always a hafnian (pfaffian), with suitable elements.

6. CONNECTION BETWEEN MATRIX ELEMENTS AND PROPAGATORS

The general matrix element $M_{fi} \equiv \langle f | U | i \rangle$ between arbitrary initial states $|i\rangle$ and final states $|f\rangle$ can be shown to obtain from the Fock space wave functions $\Phi_i(y_1, \ldots, y_N | t_1, \ldots, t_{P_0})$ and $\Phi_f(x_1, \ldots, x_N | t_1, \ldots, t_{P_0})$ as follows.

$$M_{fi}(T_1, T_0) \equiv M_{fi} = C_{fi} \, \hat{\int} \, \bar{\Phi}_f \, K \begin{pmatrix} x_1 \cdots x_{N_0} \\ y_1 \cdots y_{N_0} \end{pmatrix} \begin{vmatrix} t_1 \cdots t_{P_0} \end{vmatrix} \Phi_i$$

$$d^4 x_1 \cdots \cdots d^4 x_{N_0} d^4 y_1 \cdots \cdots d^4 y_{N_0} d^4 t_1 \cdots \cdots d^4 t_{P_0},$$

$$\tag{3.35}$$

where T_0 and T_1 denote initial and final times, respectively; N_0 is the number of electrons destroyed plus the number of positrons created; P_0 is the number of photons destroyed plus the number of photons created; C_{fi} is a *real* numerical coefficient; and $\bar{\Phi} \equiv \Phi^* \gamma^4$ for all fermions.

The sign $\hat{\int}$ has the following significance. We first integrate over the spatial arguments of all wave functions taken at times $T < T_0$ for the initial and $T' > T_1$ for the final particles. Then we average over the time interval $(T_0 - T)$ for the first case and over $(T' - T_1)$ for the second case. Thus we have, for *ingoing* particles,

$$\hat{f} = \lim \frac{1}{T_0 - T} \int_T^{T_0} dx^0 \, d^3\mathbf{x} \tag{3.36a}$$

and for *outgoing* particles

$$\hat{f} = \lim \frac{1}{T' - T_1} \int_{T_1}^{T'} dx^0 \, d^3\mathbf{x} \tag{3.36b}$$

The *propagator* $K = K_{N_0 P_0}$ introduced in (3.35) is defined by

$$K_{N_0 P_0} = \sum_{N(P_0)} \frac{\lambda^N}{N!} \int d\xi_1 \cdots \int d\xi_N \gamma' \cdots \gamma^N$$

$$\times \begin{pmatrix} x_1 \cdots x_{N_0} \, \xi_1 \cdots \xi_N \\ y_1 \cdots y_{N_0} \, \xi_1 \cdots \xi_N \end{pmatrix} [t_1 \cdots t_{P_0} \, \xi_1 \cdots \xi_N] \tag{3.37}$$

where $N + P_0 =$ even integer and $\Sigma_{N(P_0)}$ means summing over all N that have the same parity as P_0. Furthermore

$$\gamma^1 \equiv \gamma^{\mu_1}_{\alpha_1 \beta_1}, \text{ etc.,}$$

and the integrations are carried out over a finite or infinite space-time volume Ω. The elements of the determinant

$$\begin{pmatrix} x_1 \cdots \xi_N \\ y_1 \cdots \xi_N \end{pmatrix}$$

are defined by

$$(xy) = \tfrac{1}{2} S^F_{\beta\alpha}(x - y); \qquad (\gamma \partial_x + m_f)(xy) = i\delta(x - y)$$

and those of the hafnian by

$$[xy] = \tfrac{1}{2} D^F_{\beta\alpha}(x - y); \qquad (\Box_x - m_b{}^2)[xy] = i\delta(x - y).$$

The expression for $K_{N_0 P_0}$ in (3.37) gives the contributions of all Feynman graphs that possess N_0 external fermion lines and P_0 external boson lines.

Similar results can of course be written on inspection for any other field-theoretical expansion: pfaffians or determinants come from fermion fields; hafnians from boson fields. Their striking formal simplicity is a consequence of our use of (a) propagators, and (b) x space.

As is well known, the connection between propagators and Green's functions is given by

$$G\begin{pmatrix} x_1 \cdots x_{N_0} \\ y_1 \cdots y_{N_0} \end{pmatrix} t_1 \cdots t_{P_0} = G_{N_0 P_0} = \frac{K_{N_0 P_0}}{K_{00}}; \qquad (3.38)$$

the perturbative expansions of Green's functions have far more compli-
cated structures.

It is a remarkable fact that in configuration space a propagator's nth-
order perturbative term splits conveniently into boson and fermion
parts. Given *any* interaction Lagrangian, the corresponding perturbative
expansion can be written at once: any T product of free boson fields
yields a hafnian, of free fermion fields a pfaffian (which degenerates into
a determinant unless it is a Majorana field).

Part II
Equations for Propagators and Perturbative Expansions

Chapter 4

General Remarks

1. INTRODUCTION

In this part we shall be concerned mostly with the "formal" theory of propagators and Green's functions [17–22]. By this we mean the study of the equations that, in a given field theory, connect them all, the perturbative expansions being their formal solutions. As they stand, all such expressions are *meaningless,* because they contain ill-defined or divergent quantities. This problem will be taken up in Part III, where it will be shown that appropriate techniques (i.e., a suitable type of "renormalization") make them meaningful both mathematically and physically; it will be seen indeed to be a main feature of our approach to renormalization that it leaves the *form* of all such expressions and equations *invariant* (save for the addition of specific rules to handle integrations over products of distributions): requiring this invariance will be shown to be identical with requiring that *unitarity* and *causality* be preserved by renormalization.

We shall therefore be able to use all the material of the previous parts and of this part for handling the correctly renormalized theories (e.g., in approximated computations). In particular, the use of our combinatoric methods and equations will greatly simplify all of Part III, since it allows us to forgo the need for graph-by-graph analyses.

The formal study of the present part becomes mathematically correct if our expressions are somehow *regularized,* for example, by taking a *finite* space–time volume of integration Ω and only a *finite* number of fermion and boson states in the free fields (or some other sort of cutoff). It is in any case instructive to devote some attention to it, because its *algebraic* part is the same as for the exact theory, and can give therefore useful insights.

Taking, as we do, the complete set of equations for propagators *as the definition of a field theory* is, of course, a different approach from that offered by axiomatic field theory; it is worth investigating for this very reason, and a future comparison between the two methods may prove useful. In our formulation, *crossing* and *Lorentz invariance* are obvious throughout, as well as *locality* (troublesome terms will be renormalized away, and involve at most a few derivatives of a δ-function); *unitarity* also offers no problem, as we shall see next. The key problem is the *existence* of solutions (which it may prove possible to demonstrate in general) with *acceptable physical features* (e.g. positive masses); a number of "truncated" theories (i.e., models) can be proved to satisfy these requirements, but there is still a long way ahead.

2. UNITARITY

The elements of the S matrix (or U matrix, for that matter) depend essentially on two different data: the *specific particle states* between which an element is taken, and the *number of particles* involved. It turns out that only the latter determines the formal properties that propagators must satisfy in order that the S or U matrix be unitary. The elements of the U matrix $U(t_2, t_1)$ are given, we recall, by (3.35):

$$M_{fi}(t_2, t_1) = \langle f \mid K^{t_2}_{t_1} \mid i \rangle$$

$$= c_{fi} \, \hat{f} \overline{\Phi}_f K^{t_2}_{t_1} \begin{pmatrix} x_1 \cdots x_{N_0} \\ y_1 \cdots y_{N_0} \end{pmatrix} t_1 \cdots t_{P_0} \Big) \, \Phi_i, \tag{4.1}$$

where c_{fi} is a *real* normalization coefficient; $K^{t_2}_{t_1}$ is the propagator, from t_1 to t_2, for either electrodynamics or meson dynamics; and $\hat{f}$ denotes integration over all space coordinates and the average over all times $T < t_1$ for initial particles, and $T' > t_2$ for final particles.

Unitarity of the U matrix requires that

$$[U(t_2, t_1)]^\dagger = [U(t_2, t_1)]^{-1}. \tag{4.2}$$

In addition to relation (4.2) the U matrix must also satisfy the *group property* (also called macroscopic causality, i.e., causality in the large), expressed by

$$U(t_3, t_2)U(t_2, t_1) = U(t_3, t_1) \tag{4.3}$$

and

$$U(t_2, t_1)U(t_1, t_2) = 1. \tag{4.4}$$

We shall assume that the group property (4.3) holds in any theory in which the elements M_{fi} of the U matrix can be computed from formula (4.1) with the aid of propagators [23]. (If we take as the rule of our game that propagators are known only as solutions of some equations or through their series expansions, this fact must be proved by actual computation; this is an exercise in combinatorics (cf. [23, Appendix].) We then obtain from Eq. (4.4)

$$[U(t_2, t_1)]^\dagger = U(t_1, t_2) \tag{4.5a}$$

or

$$\langle f \mid K_{t_1}^{t_2} \mid i \rangle^\dagger = \langle i \mid K_{t_2}^{t_1} \mid f \rangle, \tag{4.5b}$$

which says that the unitarity condition (4.5a) will be satisfied if and only if

$$K_{t_1}^{t_2}\left(\begin{matrix} x_1 \cdots x_{N_0} \\ y_1 \cdots y_{N_0} \end{matrix} \middle| t_1 \cdots t_{P_0}\right) = (\Pi\gamma^4)$$

$$\cdot \left[K_{t_2}^{t_1}\left(\begin{matrix} y_1 \cdots y_{N_0} \\ x_1 \cdots x_{N_0} \end{matrix} \middle| t_1 \cdots t_{P_0}\right)\right]^* (\Pi\gamma^4), \tag{4.6}$$

where the products $(\Pi\gamma^4)$ arise from our particular normalization $(\bar{\psi} = \psi^+\gamma^4)$.

Example. Consider the free massless boson propagator between times t_1 and t_2

$$[xy]_{m=0} |_{t_1}^{t_2} = \{(2\pi)^2 [(x - y)^2 + i\epsilon]\}^{-1}, \qquad \epsilon > 0. \tag{4.7a}$$

By eq. (4.6) this expression must be equal to

$$([yx]_{m=0} |_{t_2}^{t_1})^* = \frac{1}{(2\pi)^2}\left[\frac{1}{(y - x)^2 - i\epsilon}\right]^*; \tag{4.7b}$$

hence

$$[xy]_{m=0} |_{t_1}^{t_2} = [yx]_{m=0}^* |_{t_2}^{t_1} \tag{4.8}$$

(where the $+i\epsilon$ in (4.7a), i.e., in $K_{t_1}^{t_2}$, implies causality from past to future, while the $-i\epsilon$ in $[(y - x)^2 - i\epsilon]^{-1}$, i.e., in $K_{t_2}^{t_1}$ implies causality from future to past). A similar result holds for *massive* free boson and fermion propagators. It appears that the requirement of unitarity becomes quite trivial when imposed on propagators.

3. A REMARK ABOUT FEYNMAN GRAPHS

Are Feynman diagrams [17–21] a realistic way of describing nature?
It seems that, at least with fermions, the answer to this question is yes, if
we consider *all* Feynman diagrams of a given order of a perturbative
expansion (regarded as asymptotic); no, if we take only a restricted sub-
set of permissible diagrams [24]. To examine this problem, let us consider a fermion field with a finite number of free states, or modes, F.
For N_0 initial and N_0 final fermions and *no* bosons present ($P_0 = 0$), the
propagator $k_{N_0 P_0}$ in Eq. (3.37) reduces to

$$K_{N_0 0} \equiv K_{N_0} \begin{pmatrix} x_1 & \cdots & x_{N_0} \\ y_1 & \cdots & y_{N_0} \end{pmatrix}$$

$$= \sum_{n=0}^{\infty} \frac{\lambda^{2n}}{(2n)!} \int_{t_i}^{t_f} d\xi_1 \cdots \int_{t_i}^{t_f} d\xi_{2n} \, \gamma^1 \cdots \gamma^{2n}$$

$$\begin{pmatrix} x_1 \cdots x_{N_0} & \xi_1 \cdots \xi_{2n} \\ y_1 \cdots y_{N_0} & \xi_1 \cdots \xi_{2n} \end{pmatrix} [\xi_1 \xi_2 \cdots \xi_{2n}].$$

We take this opportunity to remark that this expression can be simplified
to *either*

$$K_{N_0 0} = \sum_{n=0}^{\infty} \left(\frac{\lambda^2}{2}\right)^n \frac{1}{n!} \int \begin{pmatrix} x_1 \cdots x_{N_0} & \xi_1 \cdots \xi_{2n} \\ y_1 \cdots y_{N_0} & \xi_1 \cdots \xi_{2n} \end{pmatrix}$$

$$[\xi_1 \xi_2][\xi_3 \xi_4] \cdots [\xi_{2n-1} \xi_{2n}] \, d\xi_1 \cdots d\xi_{2n}, \tag{4.9}$$

where $[xy]$ and (xy) are, respectively, the free causal boson and fermion
propagators; *or*

$$K_{N_0 0} = \sum_{n=0}^{\infty} \lambda^{2n} \int_{t_i}^{t_f} d\xi_1 \int_{t_i}^{t_1} d\xi_2 \cdots \int_{t_i}^{t_{2n-1}} d\xi_{2n}$$

$$\begin{pmatrix} x_1 \cdots x_{N_0} & \xi_1 \cdots \xi_{2n} \\ y_1 \cdots y_{N_0} & \xi_1 \cdots \xi_{2n} \end{pmatrix} [\xi_1 \xi_2 \cdots \xi_{2n}] \tag{4.10}$$

where t_i and t_f are initial and final times, respectively; $d\xi_i \equiv d^4 \xi_i$ as usual;
and the spatial integration limits are understood. The determinant

$$\begin{pmatrix} x_1 \cdots \xi_{2n} \\ y_1 \cdots \xi_{2n} \end{pmatrix}$$

contains the fermion propagators S^F, while the boson propagators D^F are contained in $[\cdots]$.

Let us consider (4.10); in order to simplify the algebra, we shall work with the vacuum–vacuum transition amplitude, in which case the determinant in it becomes

$$D \equiv \begin{pmatrix} \xi_1 \cdots \xi_{2n} \\ \xi_1 \cdots \xi_{2n} \end{pmatrix} \tag{4.11}$$

with elements given by (recall that $F \equiv$ number of fermion modes):

$$(\xi_h \xi_k) = \sum_{j=1}^{F} V_{\alpha_h}(p_j) V_{\beta_k}(p_j) \exp\left[ip_j(\xi_h - \xi_k)\right] \qquad \text{for} \quad \xi_h{}^0 < \xi_k{}^0, \tag{4.12a}$$

$$(\xi_h \xi_k) = -\sum_{j=1}^{F} U_{\beta_k}(p_j) U_{\alpha_h}(p_j) \exp\left[-ip_j(\xi_h - \xi_k)\right] \qquad \text{for} \quad \xi_h{}^0 > \xi_k{}^0. \tag{4.12b}$$

The right-hand sides of (4.12a) and (4.12b) describe, respectively, the electron and positron propagators.

For $F = 1$ (i.e., only one state for the particle and the corresponding state for the antiparticle), the determinant D can be expanded by the elements of the first two rows, yielding

$$D = \sum_{C_i} (-1)^{P_i} \begin{pmatrix} \xi_1 \xi_2 \\ \xi_{i_1} \xi_{i_2} \end{pmatrix} \begin{pmatrix} \xi_3 \cdots \xi_{2n} \\ \xi_{i_3} \cdots \xi_{i_{2n}} \end{pmatrix}, \tag{4.13}$$

$$D = \sum_{C_i(i_1 > 2)} (-1)^{P_i} \begin{pmatrix} \xi_1 \xi_2 \\ \xi_{i_1} \xi_{i_2} \end{pmatrix} \begin{pmatrix} \xi_3 \xi_4 \cdots \xi_n \\ \xi_1 \xi_2 \xi_{i_5} \cdots \xi_{i_{2n}} \end{pmatrix} + \cdots, \tag{4.14}$$

P_i being the parity of the combination C_i of the indices $i_i < i_2; i_3 < \cdots < i_{2n}$. It now follows from Eq. (4.12) and the time sequence imposed by (4.10) that *all minors* with $2 < i_1 < i_2$ vanish identically ($F = 1$):

$$\begin{pmatrix} \xi_1 \xi_2 \\ \xi_{i_1} \xi_{i_2} \end{pmatrix} = \begin{vmatrix} (\xi_1 \xi_{i_1}) & (\xi_1 \xi_{i_2}) \\ (\xi_2 \xi_{i_1}) & (\xi_2 \xi_{i_2}) \end{vmatrix} = 0, \tag{4.15a}$$

so that the general determinant D can be brought to having a certain number of zeros above its main diagonal. It can likewise be shown that for F *finite*, the determinant D becomes emptier and emptier as its order increases (i.e., as more and more zeros appear). This means that there occur more and more *cancellations among graphs of different topology* as the order $2n$ of the term increases.

Example.   Consider Eq. (4.14) for the case $n = 2$ (F still equals 1). The terms, which vanish in (4.14) because of (4.15), reduce to

$$\begin{pmatrix} \xi_1 \xi_2 \\ \xi_3 \xi_4 \end{pmatrix} \begin{pmatrix} \xi_3 \xi_4 \\ \xi_1 \xi_2 \end{pmatrix} = \left\{ \begin{pmatrix} \xi_1 \xi_2 \\ \xi_3 \xi_4 \end{pmatrix} (\xi_3 \xi_1)(\xi_4 \xi_2) \right\}$$

$$- \left\{ \begin{pmatrix} \xi_1 \xi_2 \\ \xi_3 \xi_4 \end{pmatrix} (\xi_3 \xi_2)(\xi_4 \xi_1) \right\}. \tag{4.16}$$

Each of the two terms $\{\cdots\}$ vanishes by itself. This result is shown schematically in the accompanying diagram, where the four-corner loop on the left-hand side is canceled by the two disconnected two-corner loops on the right (photon lines are irrelevant).

$\xi_1{}^0$
$\xi_2{}^0$
$\xi_3{}^0$
$\xi_4{}^0$

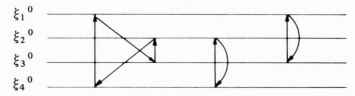

Of course, when F is *infinite,* it is *not* generally true that some Feynmann graphs are completely canceled by other graphs of a different topological structure; but it should be evident that there are tremendous interferences.

4. CONVERGENCE OF THE FERMION PERTURBATIVE EXPANSION

We shall demonstrate that all the regularized perturbative expansions of the propagator K_{00} of quantum electrodynamics converge, the radius of convergence being *finite* [25, 26]. To this effect we start with the propagator (3.37) for P_0 bosons and N_0 fermions:

$$K_{N_0 P_0} = \sum_{N(P_0)} \frac{\lambda^N}{N!} \int d\xi_1 \cdots d\xi_N \gamma^1 \cdots \gamma^N$$

$$\begin{pmatrix} x_1 \cdots x_{N_0} \, \xi_1 \cdots \xi_N \\ y_1 \cdots y_{N_0} \, \xi_1 \cdots \xi_N \end{pmatrix} [t_1 \cdots t_{P_0} \, \xi_1 \cdots \xi_N], \tag{4.17}$$

the integrations being carried out over a finite space–time volume Ω. Under the assumption that the fermion and boson propagators are bounded, in modulus, by

$$|(xy)| < M_F \qquad \text{and} \qquad |[xy]| < M_B, \tag{4.18}$$

we find immediately the following upper bound for $K_{N_0 P_0}$:

$$|K_{N_0 P_0}| < \sum_{N(P_0) = 0}^{\infty} \frac{\lambda^N}{N!}\, \Omega^N (4^3)^N (N + P_0 - 1)!!\, M_B^{(N + P_0)/2}$$

$$M_F^{N + N_0}(N + N_0)^{(N + N_0)/2}. \tag{4.19}$$

In the absence of any *external* bosons ($P_0 = 0$) and fermions ($N_0 = 0$), the last inequality reduces to

$$|K_{00}| < \sum_{n = 0}^{\infty} \frac{\lambda^{2n}}{(2n)!}\, \Omega^{2n}(4^3)^{2n}(2n - 1)!!\, M_B{}^n M_F^{2n}(2n)^n,$$

or, remembering that $(2n - 1)!! = (2n)!/2^n n!$, we have

$$|K_{00}| < \sum_{n = 0}^{\infty} \frac{\lambda^{2n}}{n!}\, \Omega^{2n} M_B{}^n M_F{}^{2n} 4^{6n} n^n. \tag{4.20}$$

In deriving this result we have used *Hadamard's inequality*, which states that the absolute value of an $N \times N$ determinant D is given by

$$|\det D| \leqslant A^N N^{N/2} \tag{4.21}$$

where A is the maximum value of each element in D.

It is clear from (4.20) that the vacuum–vacuum propagator K_{00} has a *finite* radius of convergence. Using a more elaborate procedure, based upon the considerations of the previous section, it is possible to show that if the cutoff is obtained by taking a finite number F of electron and positron states (but *not* otherwise), then the radius of convergence is actually *infinite*. This fact should not raise any hopes, however, because we shall find, in a very simple model, that *after renormalization* the corresponding propagator is *not holomorphic* for $\lambda = 0$.

Chapter 5

Formal Theory

1. EQUATIONS FOR PROPAGATORS

In this chapter we shall give formulas that are also basic in renormalization and similar work involving Green's functions. To begin with, we shall write down a complete set of integral equations connecting propagators and a corresponding set for Green's functions. Our system of equations will be in the form of recursive relations, each equation containing only a finite number of terms.

It is important to distinguish two types of equations [1]: those of *type I*, which are the same as those by Lehmann, Symanzik, and Zimmermann [27], and those of *type II*, which express the derivatives of the propagators with respect to charges and masses.

Beginning our discussion with equations of type II, we recall that the propagator for N_0 external fermions and P_0 external bosons is given by

$$K\binom{x_1 \cdots x_{N_0}}{y_1 \cdots y_{N_0}}\bigg| t_1 \cdots t_{P_0}\bigg) = \sum_{N+P_0=\text{even}} \frac{\lambda^N}{N!} \int d\xi_1 \cdots \int d\xi_N$$

$$\sum \gamma^1 \cdots \gamma^N \binom{x_1 \cdots x_{N_0} \xi_1 \cdots \xi_N}{y_1 \cdots y_{N_0} \xi_1 \cdots \xi_N}$$

$$[t_1 \cdots t_{P_0} \xi_1 \cdots \xi_N]; \tag{5.1}$$

(xy) is the causal solution of

$$(\gamma \, \partial_x + m_f)(xy) = i\delta(x - y) \tag{5.2}$$

and $[xy]$ of

$$(\Box_x - m_b{}^2)[xy] = i\delta(x - y), \tag{5.3}$$

m_f being the fermion mass, m_b the boson mass; spinor and vector indices are suppressed. Let us first compute the derivatives of (xy) and $[xy]$

with respect to m_f and $m_b{}^2$, respectively. For the fermion propagator we obtain, using Eq. (5.2),

$$\frac{\partial(xy)}{\partial m_f} = i \int d\xi (x\xi)(\xi y) = -i \int d\xi \begin{pmatrix} \overset{\circ}{\xi} x \\ \underset{\circ}{\xi} y \end{pmatrix} \tag{5.4}$$

where

$$(\overset{\circ\circ}{\xi\xi}) = 0, \qquad (\overset{\circ}{\xi}y) = (\xi y), \qquad (x\overset{\circ}{\xi}) = (x\xi); \tag{5.5}$$

$$\frac{\partial[xy]}{\partial m_b{}^2} = -i \int d\xi [x\xi][\xi y] = -\frac{i}{2} \int d\xi [\overset{\circ\circ}{\xi\xi} xy]; \tag{5.6}$$

or, in general,

$$\frac{\partial}{\partial m_b{}^2} [\alpha_1 \cdots \alpha_{2n}] = \sum_{h<k} \frac{\partial[\alpha_h \alpha_k]}{\partial(m_b{}^2)} \frac{\partial[\alpha_1 \cdots \alpha_{2n}]}{\partial[\alpha_h \alpha_k]}$$

$$= -\frac{i}{2} \int d\xi [\overset{\circ\circ}{\xi\xi} \alpha_1 \cdots \alpha_{2n}], \tag{5.7}$$

where the elements of the hafnian are functions of the mass m_b. With the foregoing $\overset{\circ}{\xi}$ notation, we have the following concise differentiation rule for determinants:

$$\frac{\partial}{\partial m_f} \begin{pmatrix} \alpha_1 \cdots \alpha_n \\ \beta_1 \cdots \beta_n \end{pmatrix} = \sum_{h,k=1}^{n} \frac{\partial(\alpha_h \alpha_k)}{\partial m_f} \frac{\partial \begin{pmatrix} \alpha_1 \cdots \alpha_n \\ \beta_1 \cdots \beta_n \end{pmatrix}}{\partial(\alpha_h \alpha_k)}$$

$$= i \sum_{h,k} \int d\xi (\alpha_h \xi)(\xi\beta_k)(\mathrm{Adj})_k{}^h$$

$$= -i \int d\xi \begin{pmatrix} \overset{\circ}{\xi} \alpha_1 \cdots d_n \\ \underset{\circ}{\xi} \beta_1 \cdots \beta_n \end{pmatrix}. \tag{5.8}$$

From (5.1) and (5.7) we derive the first important equations:

$$\frac{\partial}{\partial m_f} K_{N_0 P_0} = -i \int d\xi K \begin{pmatrix} \overset{\circ}{\xi} x_1 \cdots x_{N_0} \\ \underset{\circ}{\xi} y_1 \cdots y_{N_0} \end{pmatrix} t_1 \cdots t_{P_0} \end{pmatrix}; \tag{5.9}$$

$$\frac{\partial}{\partial m_b{}^2} K_{N_0 P_0} = -\frac{i}{2} \int d\xi K \begin{pmatrix} x_1 \cdots x_{N_0} \\ y_1 \cdots y_{N_0} \end{pmatrix} \overset{\circ\circ}{\xi\xi} t_1 \cdots t_{P_0} \end{pmatrix}. \tag{5.10}$$

The remaining variable λ yields

$$\frac{\partial}{\partial \lambda} K_{N_0 P_0} = \int d\xi \sum \gamma^\xi K \begin{pmatrix} x_1 \cdots x_{N_0} \xi \\ y_1 \cdots y_{N_0} \xi \end{pmatrix} \xi t_1 \cdots t_{P_0} \end{pmatrix}. \tag{5.11}$$

Equations (5.9), (5.10), and (5.11) are the *branching equations* involving the derivatives of the propagators $K_{N_0 P_0}$. The differentiation in (5.9) implies physically that another fermion line is being added; in (5.10) that another boson line is added.

The equations (5.9) and (5.10) cannot be iterated as given, because the ξ and $\overset{\circ}{\xi}$ terms interfere. To solve this problem, we shall rewrite these equations in terms of ξ only:

$$\frac{\partial K_{N_0 P_0}}{\partial m_f} = -i \int d\xi \, K\left(\begin{matrix} \xi \, x_1 \cdots x_{N_0} \\ \xi \, y_1 \cdots y_{N_0} \end{matrix}\middle| t_1 \cdots t_{P_0}\right)$$

$$+ i \int d\xi \, (\xi\xi) K\left(\begin{matrix} x_1 \cdots x_{N_0} \\ y_1 \cdots y_{N_0} \end{matrix}\middle| t_1 \cdots t_{P_0}\right); \tag{5.9'}$$

$$\frac{\partial K_{N_0 P_0}}{\partial m_b{}^2} = -\frac{i}{2} \int d\xi \, K\left(\begin{matrix} x_1 \cdots x_{N_0} \\ y_1 \cdots y_{N_0} \end{matrix}\middle| \xi\xi \, t_1 \cdots t_{P_0}\right)$$

$$+ \frac{i}{2} \int d\xi \, [\xi\xi] K\left(\begin{matrix} x_1 \cdots x_{N_0} \\ y_1 \cdots y_{N_0} \end{matrix}\middle| t_1 \cdots t_{P_0}\right) \tag{5.10'}$$

and

$$\frac{\partial K_{N_0 P_0}}{\partial \lambda} = \int d\xi \, \Sigma \, \gamma^\xi K\left(\begin{matrix} x_1 \cdots x_{N_0} \, \xi \\ y_1 \cdots y_{N_0} \, \xi \end{matrix}\middle| \xi \, t_1 \cdots t_{P_0}\right). \tag{5.11'}$$

The last equation is identical with (5.11).

Let us now call

$$(\xi\xi) = f(\xi, m_f), \tag{5.12a}$$

$$[\xi\xi] = g(\xi, m_b{}^2), \tag{5.12b}$$

where f and g are some differentiable functions of the corresponding masses (and of position if there are external fields), and write $K_{N_0 P_0}$ in the form [28, 29]

$$K_{N_0 P_0} = \bar{K}_{N_0 P_0} \exp[+i \int d\xi \int^{m_f} dm' f(\xi, m')$$

$$+ \frac{i}{2} \int^{m_b{}^2} d(m')^2 g(\xi, (m')^2)] \tag{5.13}$$

(i.e., we *define* $\bar{K} = K \exp(-[\cdots])$). The last equation holds in both quantum electrodynamics and meson dynamics. Equations (5.9'),

$(5.10')$, and $(5.11')$ can now be written strictly in terms of $\tilde{K}$:

$$\frac{\partial}{\partial m_f} \tilde{K}\begin{pmatrix} x_1 \cdots x_{N_0} \\ y_1 \cdots y_{N_0} \end{pmatrix} t_1 \cdots t_{P_0} \bigg) = -i \int d\xi\, \tilde{K}\begin{pmatrix} x_1 \cdots x_{N_0}\, \xi \\ y_1 \cdots y_{N_0}\, \xi \end{pmatrix} t_1 \cdots t_{P_0} \bigg),$$

$$(5.14)$$

$$\frac{\partial}{\partial m_b{}^2} \tilde{K}\begin{pmatrix} x_1 \cdots x_{N_0} \\ y_1 \cdots y_{N_0} \end{pmatrix} t_1 \cdots t_{P_0} \bigg) = -\frac{i}{2} \int d\xi\, \tilde{K}\begin{pmatrix} x_1 \cdots x_{N_0} \\ y_1 \cdots y_{N_0} \end{pmatrix} t_1 \cdots t_{P_0}\, \xi\xi \bigg),$$

$$(5.15)$$

and obviously

$$\frac{\partial}{\partial \lambda} \tilde{K}\begin{pmatrix} x_1 \cdots x_{N_0} \\ y_1 \cdots y_{N_0} \end{pmatrix} t_1 \cdots t_{P_0} \bigg) = \int d\xi\, \Sigma\, \gamma^\xi\, \tilde{K}\begin{pmatrix} x_1 \cdots x_{N_0}\, \xi \\ y_1 \cdots y_{N_0}\, \xi \end{pmatrix} t_1 \cdots t_{P_0}\, \xi \bigg),$$

$$(5.16)$$

where the subscript $N_0 P_0$ in $\tilde{K}$ has been suppressed. The last three linear equations are what we call *type II equations*; they clearly hold for propagators, but not for Green's functions, which are defined by

$$G_{N_0 P_0} = \frac{K_{N_0 P_0}}{K_{00}}.$$

We call branching equations of the *first type* the following (which can be derived in many ways) [1, 27, 30, etc.]

$$\tilde{K}_{N_0 P_0}\begin{pmatrix} x_1 \cdots x_{N_0} \\ y_1 \cdots y_{N_0} \end{pmatrix} t_1 \cdots t_{P_0} \bigg) = \sum_{k=2}^{P_0} [t_1 t_h] \tilde{K}\begin{pmatrix} x_1 \cdots x_{N_0} \\ y_1 \cdots y_{N_0} \end{pmatrix} t_2 \cdots$$

$$t_{h-1} t_{h+1} \cdots t_{P_0} \bigg)$$

$$+ \lambda \int d\xi\, \Sigma\, \gamma^\xi [t_1 \xi] \tilde{K}\begin{pmatrix} x_1 \cdots x_{N_0}\, \xi \\ y_1 \cdots y_{N_0}\, \xi \end{pmatrix} t_2 \cdots t_{P_0} \bigg),$$

$$(5.17)$$

which shows that we can *decrease* the number of external boson lines P_0. Similarly, the following equation enables us to extract *fermion* lines.

$$\tilde{K}_{N_0 P_0} = \sum_{h=1}^{N_0} (-1)^{h-1} (x_1 y_h) \tilde{K}\begin{pmatrix} x_2 \cdots x_{N_0} \\ y_1 \cdots y_{h-1}\, y_{h+1} \cdots y_{N_0} \end{pmatrix} t_1 \cdots t_{P_0} \bigg)$$

$$- \lambda \int d\xi\, \Sigma\, \gamma^\xi (x_1 \xi) \tilde{K}\begin{pmatrix} \xi\, x_2 \cdots x_{N_0} \\ y_1 y_2 \cdots y_{N_0} \end{pmatrix} \xi\, t_1 \cdots t_{P_0} \bigg).$$

If we are interested only in *purely fermionic* propagators, we set $P_0 = 0$ to obtain

$$\bar{K}_{N_0 0}\begin{pmatrix} x_1 \cdots x_{N_0} \\ y_1 \cdots y_{N_0} \end{pmatrix}$$

$$= \sum_{h=1}^{N_0} (-1)^{h-1}(x_1 y_h)\bar{K}\begin{pmatrix} & x_2 \cdots x_{N_0} & \\ y_1 \cdots y_{h-1} y_{h+1} \cdots y_{N_0} & \bigg| t_1 \cdots t_{P_0} \end{pmatrix}$$

$$- \lambda^2 \int d\xi_1 \int d\xi_2 \sum \gamma^{\xi_1}\gamma^{\xi_2}(x_1\xi_1)[\xi_1\xi_2]\bar{K}_{(N_0+1)0}\begin{pmatrix} \xi_1 \ x_2 \cdots x_{N_0} \ \xi_2 \\ y_1 \cdots y_{N_0} \ \xi_2 \end{pmatrix}.$$

$$(5.19)$$

The type I equations coincide with those discussed in considerable detail by Lehmann *et al.* [27].

We note that in the limit as $\lambda \to 0$ all the equations derived thus far, except (5.16), reduce to relations that relate correctly *free* propagators for arbitrary numbers of fermions and bosons; Eqs. (5.16) give all λ derivatives of all propagators at $\lambda = 0$. Therefore, to assume the *whole* set of branching equations as the basis of our field theory implies that we also assume the existence (after renormalization) of the derivatives to *all* orders of propagators and Green's functions (but *not* the convergence of perturbative expansions, which may be only asymptotic). In practice, this means that manipulations performed on the *formal* perturbative expansions can be used as shorthand for equivalent manipulations on the branching equations, and conversely. This will be seen to be true also after our renormalization procedure is applied, within limits: *great care* will then be necessary to make sure that it is indeed so.

It is important to remark that, if only equations of type I, such as (5.19), are used *and no other conditions imposed,* their solutions are clearly *not unique*: we can assign arbitrarily K_{00} and K_{20}, and then deduce from (5.19) a special value of K_{30}, and so on, with increasing degrees of arbitrariness. This will be most apparent in a numerical model of the theory which is discussed in Section 4 of this chapter. Inclusion of type II equations may eliminate this arbitrariness.

2. PERTURBATIVE EXPANSIONS FOR GREEN'S FUNCTIONS; THE LINKED-CLUSTER EXPANSION

Up to this point we have been concerned almost exclusively with the perturbative expansions of propagators. In this section we shall derive perturbative expansions for Green's functions. We shall work with purely fermionic propagators in order to simplify algebraic manipulations as much as possible. Thus we have, from (5.1) and using $P_0 = 0 (G_{N_0 P_0} \equiv K_{N_0 P_0}/K_{00})$,

$$
G_{N_0 0} = G \begin{pmatrix} x_1 & \cdots & x_{N_0} \\ y_1 & \cdots & y_{N_0} \end{pmatrix} = \frac{1}{K_{00}} \sum_{n=0}^{\infty} \left(\frac{\lambda^2}{2} \right)^n \frac{1}{n!} \int \Sigma \gamma^1 \cdots
$$
$$
\gamma^{2n} \begin{pmatrix} x_1 & \cdots & x_{N_0} & \xi_1 & \cdots & \xi_{2n} \\ y_1 & \cdots & y_{N_0} & \xi_1 & \cdots & \xi_{2n} \end{pmatrix} [\xi_1 \xi_2] \cdots [\xi_{2n-1} \xi_{2n}];
$$
$$
\tag{5.20}
$$

Eq. (5.19), on the other hand, reduces to

$$
G_{N_0 0} \begin{pmatrix} x_1 & \cdots & x_{N_0} \\ y_1 & \cdots & y_{N_0} \end{pmatrix} = \sum_{h=1}^{N_0} (-1)^{h-1} (x_1 y_h) G \begin{pmatrix} & x_2 & \cdots & x_{N_0} \\ y_1 & \cdots & y_{h-1} y_{h+1} & \cdots & y_{N_0} \end{pmatrix}
$$
$$
+ (-1)^{N_0} \lambda^2 \int d\xi_1 \int d\xi_2 \, \Sigma \, \gamma^{\xi_1} \gamma^{\xi_2} (x_1 \xi_1) [\xi_1 \xi_2]
$$
$$
G_{(N_0 + 1)0} \begin{pmatrix} \xi_1 x_2 & \cdots & x_{N_0} & \xi_2 \\ y_1 & \cdots & y_{N_0} & \xi_2 \end{pmatrix}.
$$
$$
\tag{5.21}
$$

Next we introduce the notation $(N_0 > 1)$

$$
\begin{pmatrix} x_1 & \cdots & x_{N_0} & |\xi_1 \xi_2| \xi_3 \xi_4| & \cdots & |\xi_{2n-1} \xi_{2n} \\ y_1 & \cdots & y_{N_0} & |\xi_1 \xi_2| \xi_3 \xi_4| & \cdots & |\xi_{2n-1} \xi_{2n} \end{pmatrix}
$$
$$
= (x_1 y_1) \begin{pmatrix} x_2 & \cdots \\ y_2 & \cdots \end{pmatrix} - (x_1 y_2) \begin{pmatrix} x_2 & \cdots \\ y_1 y_3 & \cdots \end{pmatrix} + \cdots
$$
$$
+ (-1)^{N_0} (x_1 \xi_1) \begin{pmatrix} x_2 & \cdots & x_{N_0} & \xi_1 & \xi_2 & & |\xi_{2n-1} \xi_{2n} \\ y_1 & \cdots & y_{N_0-1} & y_{N_0} \xi_2 & & \cdots & |\xi_{2n-1} \xi_{2n} \end{pmatrix}, \tag{5.22a}
$$

with the condition, for $N_0 = 1$

$$
\begin{pmatrix} x_1 & |\xi_1 \xi_2| & \cdots \\ y_1 & |\xi_1 \xi_2| & \cdots \end{pmatrix} = - (x_1 \xi_1) \begin{pmatrix} \xi_1 \xi_2| & \cdots & |\xi_{2n-1} \xi_{2n} \\ y_1 \xi_2| & \cdots & |\xi_{2n-1} \xi_{2n} \end{pmatrix} \tag{5.22b}
$$

where the various factors on the right-hand side of (5.22) reduce in the end to determinants. The symbol introduced in (5.22) contains the prescription for the systematic dropping of the vacuum–vacuum terms in the simple form exhibited by Eqs. (5.22a) and (5.22b). It is then a straightforward matter to show that Eq. (5.20) becomes

$$G_{N_0 0} = \sum_{p=0}^{\infty} \lambda^{2p} \int \sum \gamma^1 \cdots \gamma^{2p} [\xi_1 \xi_2] \cdots [\xi_{2p-1} \xi_{2p}]$$

$$\begin{pmatrix} x_1 \cdots x_{N_0} \begin{vmatrix} \xi_1 \xi_2 \end{vmatrix} \cdots \begin{vmatrix} \xi_{2p-1} \xi_{2p} \end{vmatrix} \\ y_1 \cdots y_{N_0} \begin{vmatrix} \xi_1 \xi_2 \end{vmatrix} \cdots \begin{vmatrix} \xi_{2p-1} \xi_{2p} \end{vmatrix} \end{pmatrix}, \tag{5.23}$$

which is the desired expansion and the analogue of (5.1)

Of particular interest to many-body physicists is the quantity $\log K_{00}$, which can be derived using (note that we use here K instead of $\tilde{K}$)

$$\frac{\partial}{\partial(\lambda^2/2)} K_{N_0 0} = \int d\xi_1 \int d\xi_2 \sum \gamma^{\xi_1} \gamma^{\xi_2} [\xi_1 \xi_2] K \begin{pmatrix} x_1 \cdots x_{N_0} \, \xi_1 \, \xi_2 \\ y_1 \cdots y_{N_0} \, \xi_1 \, \xi_2 \end{pmatrix}. \tag{5.24}$$

Then

$$\frac{d}{d(\lambda^2/2)} (\log K_{00}) = \frac{1}{K_{00}} \frac{d(K_{00})}{d\lambda^2} = \iint d\xi_1 \, d\xi_2 [\xi_1 \xi_2] \sum \gamma^{\xi_1} \gamma^{\xi_2} G \begin{pmatrix} \xi_1 \, \xi_2 \\ \xi_1 \, \xi_2 \end{pmatrix},$$

$$\tag{5.25}$$

so that

$$\log K_{00} = \sum_{p=1}^{\infty} \frac{\lambda^{2p}}{2p} \int \sum \gamma^1 \cdots \gamma^{2p} [\xi_1 \xi_2] \cdots [\xi_{2p-1} \xi_{2p}]$$

$$\begin{pmatrix} \xi_1 \, \xi_2 \begin{vmatrix} \end{vmatrix} \cdots \begin{vmatrix} \xi_{2p-1} \xi_{2p} \end{vmatrix} \\ \xi_1 \, \xi_2 \begin{vmatrix} \end{vmatrix} \begin{vmatrix} \xi_{2p-1} \xi_{2p} \end{vmatrix} \end{pmatrix}, \tag{5.26}$$

which was later discovered as the *linked-cluster expansion* in many-body theory [31]. This expansion appears as a simple by-product of the field-theoretical techniques employed here.

3. EQUATIONS FOR SINGLE PROPAGATORS IN THE $g\phi^4$ THEORY

A nontrivial consequence of the branching equations (5.14), (5.15), and (5.16) is the fact that if we endeavor to derive systematically from them equations which contain *only one* propagator [32], then the infinitely many conditions that can be found from the relations of that propagator with higher- or lower-order propagators reduce to a small, finite, and easily deducible set: all other conditions can be shown to be derivatives of some order of these conditions. We shall demonstrate this for the $g\phi^4$ theory, which describes the coupling of a spinless neutral meson to itself (take $\lambda = -ig$).

The perturbative expansion of the propagator in this theory with $2N_0$ external lines (only an *even* number of lines occurs in the propagators) reads

$$K_{2N_0} \equiv K(x_1 \cdots x_{2N_0})$$

$$= \sum_{N=0}^{\infty} \frac{\lambda^N}{N!} \int d\xi_1 \cdots \int d\xi_N [x_1 \cdots x_{2N_0} \xi_1 \xi_1 \xi_1 \xi_1 \xi_2 \xi_2 \xi_2 \xi_2 \cdots$$

$$\xi_N \xi_N \xi_N \xi_N], \tag{5.27}$$

where the element $[xy] = \Delta_F(xy; m)$ denotes the free boson propagator of mass m. Writing $m^2 = \mu$, we have, for the complete set of branching equations,

$$\bar{K}(x_1 \cdots x_{2N_0}) = \sum_{h=2}^{2N_0} [x_1 x_h] \bar{K}_{2N_0-2}(x_2 \cdots x_{h-1} x_{h+1} \cdots x_{2N_0})$$

$$+ 4\lambda \int d\xi_1 [x_1 \xi_1] \bar{K}_{2N_0+2}(x_2 \cdots x_{2N_0} \xi_1 \xi_1 \xi_1), \tag{5.28}$$

$$\partial_\mu \bar{K}_{2N_0}(x_1 \cdots x_{2N_0}) = -\frac{i}{2} \int \bar{K}_{2N_0+2}(x_1 \cdots x_{2N_0} \xi_1 \xi_1) \, d\xi_1 \tag{5.29}$$

and

$$\partial_\lambda \bar{K}_{2N_0}(x_1 \cdots x_{2N_0}) = \int d\xi_1 \bar{K}_{2N_0+4}(x_1 \cdots x_{2N_0} \xi_1 \xi_1 \xi_1 \xi_1). \tag{5.30}$$

The branching equation (5.28) is of type I, while (5.29) and (5.30) are of type II.

The last three formulas may be simplified by means of the following rules.

(i) The substitution of an integration variable ξ_h by the index h means that the integration over ξ_h is understood;

(ii) in a product of free propagators, the external variables inside the propagators are not written explicitly; it is understood that the external variables that appear as arguments in the free propagators are deleted from the arguments of the propagators.

Equations (5.28), (5.29), and (5.30) then simplify to

$$\bar{K}_{2N_0} = \sum_{h=2}^{2N_0} [x_1 x_h] \bar{K}_{2N_0-2} + 4\lambda [x_1 1] \bar{K}_{2N_0+2}(1\ 1\ 1), \tag{5.31}$$

$$2i \, \partial_\mu \bar{K}_{2N_0} = \bar{K}_{2N_0+2}(1\ 1), \tag{5.32}$$

$$\partial_\lambda \bar{K}_{2N_0} = \bar{K}_{2N_0+4}(1\ 1\ 1\ 1). \tag{5.33}$$

With the aid of Eqs. (5.31)-(5.33) it is possible to derive all the equations that contain only one propagator $\tilde{K}_{2N_0}$. We proceed as follows. Combining the μ and λ derivatives (5.32), (5.33), we obtain

$$\partial_\lambda \tilde{K}_{2N_0}(1\ 1\ 2\ 2) + 4\ \partial_\mu^2 \tilde{K}_{2N_0}(1\ 1\ 1\ 1) = 0. \tag{5.34}$$

The two other equations, obtained after some manipulations, are

$$2i\ \partial_\mu \tilde{K}_{2N_0}(1\ 1) = \sum_{h=2}^{2N_0-2} [x_1 x_h]\tilde{K}_{2N_0}(1\ 1\ 2\ 2)$$

$$+ 4[x_1\ 1]\tilde{K}_{2N_0}(1\ 2\ 2) - 16\lambda[x_1\ 1]\partial_\mu^2\tilde{K}_{2N_0}(1\ 1\ 1),$$

$$\tag{5.35}$$

and

$$2i\ \partial_\mu \tilde{K}_{2N_0}(1\ 1\ 1\ 1) = \sum_{h=2}^{2N_0-4} [x_1 x_h]\tilde{K}_{2N_0}(1\ 1\ 2\ 2\ 2)$$

$$+ 2[x_1\ 1]\tilde{K}_{2N_0}(1\ 2\ 2\ 2\ 2)$$

$$+ 4\partial_\lambda\{\lambda[x_1\ 1]\tilde{K}_{2N_0}(1\ 1\ 1\ 2\ 2)\}. \tag{5.36}$$

4. A NUMERICAL MODEL

In order to study the system (5.34), (5.36) further, we shall examine a simple model [33, 34] in which the free propagator is replaced by a constant which may depend on μ. Although such a model is physically meaningless, it is nevertheless useful because, retaining as it does all the algebraic characteristics of the exact theory, it sheds some light on the analytical nature we should expect for the propagators as functions of λ. With $[xy]$ = constant and working with a finite space-time volume Ω, we find that

$$[xy] = f(\mu) = \frac{i}{\Omega(a-\mu)}, \tag{5.37}$$

where $a = ic\Omega^{-1}$, c being an arbitrary constant, guarantees that (5.29) is satisfied. Equations (5.34) and (5.35) then become

$$\Omega\ \partial_\lambda \tilde{K}_{2N_0} + 4\ \partial_\mu^2\ \tilde{K}_{2N_0} = 0, \tag{5.38}$$

$$\partial_\mu^2\ \tilde{K}_{2N_0} + \frac{\Omega}{8\lambda}\ (a-\mu)\partial_\mu\tilde{K}_{2N_0} - \frac{2N_0+1}{16}\ \Omega\tilde{K}_{2N_0}. \tag{5.39}$$

It follows from Eqs. (5.38), (5.39), or directly from Eqs. (5.28)–(5.30), that one of the equations is equal to

$$\lambda^2 \frac{d^2}{d\lambda^2} \tilde{K}_{2N_0} + \left(\lambda \frac{4N_0^2 + 2N_0}{2} - \frac{1}{16\Omega f^2} \right) \frac{d}{d\lambda} \tilde{K}_{2N_0}$$
$$+ \tfrac{1}{16}(4N_0^2 + 8N_0 + 3)\tilde{K}_{2N_0} = 0 \qquad (f \equiv f(\mu)), \qquad (5.40)$$

and may be solved, at least formally, by the divergent expansion

$$\tilde{K}_{2N_0} = \sum_{N=0}^{\infty} \frac{A_N}{N!} \lambda^N$$
$$= \sum_{N=0}^{\infty} \frac{(4N + 2N_0 - 1)!!}{N!} \Omega^N f^{2N+N_0} \lambda^N. \qquad (5.41)$$

The general integral of (5.40) can be written as a function of

$$\tilde{K}_{2N_0} = A_{2N_0} H_{2N_0}(y) + B_{2N_0} I_{2N_0}(y) \qquad (5.42)$$

where

$$y = (4\lambda f^2 \Omega^2)^{-1}. \qquad (5.43)$$

Since (5.40) has an irregular singular point at $\lambda = 0$, the solution (5.42) can be expected to be singular at the same point. When we study the series (5.41) for $y \to \infty$ ($\lambda \to 0$), we find two entirely different situations:

Case i.   Re $y > 0$

The solution is *not unique:* it consists of a confluent hypergeometric function [35] of $1/\sqrt{\lambda}$, with an essential singularity at $\lambda = 0$, whose asymptotic expansion is (5.41), plus another function (not uniquely determined), which vanishes with all its derivatives for $\lambda \to 0$.

Case ii.   Re $y \leqslant 0$

The solution is *unique* and has otherwise the same properties as in case i.

We have cited this example essentially because it shows clearly *when* and *how* we can *legitimately* interchange two infinite operations to obtain a *correct convergent* solution from a formal *divergent series* [34].

Suppose we have a series like (5.41), say

$$\sum_{N=0}^{\infty} A_N \lambda^N,$$ (5.44)

which is known to be a *formal* solution of some equation, say (5.40). This means that some *finite* number of coefficients $A_{N-h} \cdots A_{N+k}$ satisfy, because of (5.40), some relations among them. This fact remains true, then, also if we replace λ^N in (5.44) with a suitable linear integral transform (or some other limiting process), and *then* exchange the Σ_N with the other *infinite* operation thus introduced: the resulting expression is also a formal solution of (5.40) and if it converges, it is a correct solution. For the sake of concreteness let us put

$$\lambda^N = \int_0^{\infty} dt \, \frac{t^{kN_0-1}}{\Gamma(kN_0)} \exp\left(-t\lambda^{-1/k}\right).$$ (5.45)

Performing on our model the exchange

$$\Sigma \int = \int \Sigma,$$ (5.46)

we find indeed, from the left-hand side of (5.41), at the right-hand side of (5.46) the convergent solutions mentioned as cases i and ii (but not, of course, the asymptotically vanishing one of case i).

In other words, two *illegitimate* operations, divergent sum and the exchange (5.46), *can be made and proved to compensate each other* to yield a correct result: the Borel summation criterion (5.45) can in this way be proved to yield the correct solution, whenever a similar situation arises.

5. NONANALYTICAL PROPERTIES OF PROPAGATORS

In the preceding section we studied a simple model which can be solved exactly and for which it was possible to derive a rigorous procedure enabling us to obtain a convergent solution from a divergent formal series expansion. The solutions of our model belong to a well-defined class of nonanalytical functions.

The most important property of analytical functions is that they are given uniquely in the whole domain of definitions $\mathscr{D}$ if their values are known in an ϵ region. There also exist other functions that are characterized in exactly the same fashion; these functions belong to the wider class of *quasi-analytical* functions and have been studied by Carleman

[36]. More generally, a function $f(\lambda)$ in $\mathscr{D}$ that satisfies

$$|f^{(n)}(\lambda)| < KA_n, \qquad n = 0, 1, 2, \ldots, \tag{5.47}$$

is said to be of class $\mathscr{A} = (A_0, A_1, A_2, \ldots)$ where $\mathscr{A}$ is called the *Hadamard class*. The functions $f(\lambda)$ are also said to be of Hadamard's class α if $A_n = (\alpha n)!$. If $A_n = n!$, the $f(\lambda)$ are clearly analytical. A sufficient condition for *quasi-analyticity* is that the series

$$\sum_{n=0}^{\infty} A_n^{-1/n}$$

be divergent [36].

The solutions of the heat equation, for example, are of *class 1* (analytical) in the space variables and of *class 2* in the time variable. Our Green's functions and propagators satisfy relation (5.47) with $A_n = (2n)!$ and are consequently of Hadamard's class 2. The same is also true of the convergent solutions of the particular model considered in Section 4. We have also found Green's functions to be of class 2 in other models: *this seems a common feature to all cases we have been able to handle.*

Chapter 6

Gauge Invariance and Infrared Divergences

1. GAUGE INVARIANCE OF GREEN'S FUNCTIONS IN QUANTUM ELECTRODYNAMICS [37–39]

With the aid of the combinatoric techniques developed in this and earlier chapters it is possible to study the gauge invariance of Green's functions [40] with a minimum of effort. The "tricks" employed here will also be useful in our later discussion of renormalization theory.

We shall study the gauge properties of the propagator

$$
\tilde{K}_{\mu_1 \cdots \mu_{P_0}}\!\left(\begin{matrix} x_1 \cdots x_{N_0} \\ y_1 \cdots y_{N_0} \end{matrix}\middle| t_1 \cdots t_{P_0}\right)
\tag{6.1}
$$

in the framework of quantum electrodynamics (qed); the $\mu_1, \ldots, \mu_P$ represent the vector indices for the external photon lines. Applications to (5.48) of

$$
\left(\gamma \frac{\partial}{\partial \xi} + m\right)(\xi y) = +i\delta(\xi - y)
\tag{6.2}
$$

yields

$$
\left(\gamma_\mu \frac{\partial}{\partial \xi_\mu}\right) \tilde{K}\!\left(\begin{matrix} \xi x_1 \cdots x_{N_0} \\ \xi y_1 \cdots y_{N_0} \end{matrix}\middle| t_1 \cdots t_{P_0}\right) \doteq -i \sum_{j=1}^{N} [\delta(x_j - \xi)
$$
$$
- \delta(y_j - \xi)] \tilde{K}\!\left(\begin{matrix} x_1 \cdots x_{N_0} \\ y_1 \cdots y_{N_0} \end{matrix}\middle| t_1 \cdots t_{P_0}\right).
\tag{6.3}
$$

For the free causal photon propagator we shall employ the form

$$
[tt'] = D_{\mu\nu}(t, t') = D_{\mu\nu}^{(0)}(t, t') + \frac{\partial^2 f(t, t')}{\partial t_\mu \, \partial t_\nu}
\tag{6.4}
$$

where $D_{\mu\nu}^{(0)}(t, t')$ is the value of $D_{\mu\nu}(t, t')$ at a fixed gauge and $f(t, t')$ is an arbitrary function. Our task is to study the transformation properties of the Green functions under a change of gauge according to (6.4); that is, we want to know what the f-dependence of the Green functions is.

To simplify the algebra, let us take purely fermionic propagators

$$\bar{K}\begin{pmatrix} x_1 \cdots x_{N_0} \\ y_1 \cdots y_{N_0} \end{pmatrix}$$

and write

$$\frac{d}{d(\lambda^2/2)} \, \bar{K}\begin{pmatrix} x_1 \cdots x_{N_0} \\ y_1 \cdots y_{N_0} \end{pmatrix} = \int d\xi_1 \int d\xi_2 \, \gamma^{\mu_1}\gamma^{\mu_2} \, D_{\mu_1\mu_2}(\xi_1, \xi_2)$$

$$\bar{K}\begin{pmatrix} x_1 \cdots x_{N_0} \, \xi_1 \xi_2 \\ y_1 \cdots y_{N_0} \, \xi_1 \xi_2 \end{pmatrix}. \tag{6.5}$$

Substituting (6.4) into the right-hand side of (6.5), we show that

$$\frac{d}{d(\lambda^2/2)} \, \bar{K}\begin{pmatrix} x_1 \cdots x_{N_0} \\ y_1 \cdots y_{N_0} \end{pmatrix} = \iint d\xi_1 \, d\xi_2 \, D^{(0)}_{\mu_1\mu_2}$$

$$\bar{K}\begin{pmatrix} x_1 \cdots x_{N_0} \, \xi_1 \xi_2 \\ y_1 \cdots y_{N_0} \, \xi_1 \xi_2 \end{pmatrix} - 2I\bar{K}\begin{pmatrix} x_1 \cdots x_{N_0} \\ y_1 \cdots y_{N_0} \end{pmatrix} \tag{6.6}$$

where

$$I \equiv \tfrac{1}{2} \sum_{i=1}^{N_0} \sum_{j=1}^{N_0} [f(x_i, x_j) + f(y_i, y_j) - f(x_i, y_j) - f(y_i, x_j)]. \tag{6.7}$$

The change in gauge, as given in (6.4), is compensated by rephasing all propagators $\bar{K}$ as follows.

$$\bar{K}(\cdots) = \exp(\lambda^2 I)\bar{K}(\cdots) \tag{6.8}$$

where I is given by (6.7); then the new propagators satisfy, in the new gauge, the same equations as the old ones. A Stückelberg field, for instance, can be immediately disposed of in this manner.

2. AN ANALYSIS OF INFRARED DIVERGENCES

A. Infrared Divergences

As is well known, unlike ultraviolet divergences, which are caused by short-range effects, infrared divergences originate from the long-range interactions among fields. We propose to show here that our approach also permits us to take care of infrared divergences, although the problem for them is not to remove, but rather to factor out of a matrix element the divergent contributions, so as to achieve an appropriate generalization of the classical results of Bloch and Nordsieck [41]: our work is thus similar in scope to that by Yennie and Suura [42].

We restrict our considerations to the case of electron–electron scattering, with the explicit exclusion of forward scattering; this will suffice, since the generalization to all possible cases is immediate and does not involve, with our method, any complication of the calculations presented here. There is in principle no interference between ultraviolet and infrared divergences, except for photon self-energies; only for the latter, indeed, can a continuous chain of short-range vacuum polarizations have an overall long-range effect in our problem. As we deal with a theory which we suppose already freed from ultraviolet infinities by renormalization, any such interference is taken automatically into account by our treatment.

We consider [43] the matrix element for nonforward scattering of two electrons with initial momenta q_1 and q_2 and final momenta p_1 and p_2. It is given by

$$M\begin{pmatrix} \overset{\circ}{p}_1 \overset{\circ}{p}_2 \\ \overset{\circ}{q}_1 \overset{\circ}{q}_2 \end{pmatrix} = \sum_{n=0}^{\infty} \frac{\lambda^{2n}}{(2n)!} \int d\xi_1 \cdots \int d\xi_{2n} \, \Sigma \, \gamma^1 \cdots \gamma^{2n}$$

$$\begin{pmatrix} \overset{\circ}{p}_1 \overset{\circ}{p}_2 \xi_1 \cdots \xi_{2n} \\ \overset{\circ}{q}_1 \overset{\circ}{q}_2 \xi_1 \cdots \xi_{2n} \end{pmatrix} [\xi_1 \cdots \xi_{2n}], \tag{6.9}$$

with

$$(\overset{\circ}{p} \, \xi) = \bar{u}_p(\xi), \quad (\xi \, \overset{\circ}{q}) = u_q(\xi), \quad (\overset{\circ}{p} \, \overset{\circ}{q}) = 0, \tag{6.10}$$

where $u_q(\xi)$ denotes the wave function of the electron of momentum q and $\bar{u} = u^* \gamma^4$.

The λ-derivative equation for propagators (cf. Eq. (5.16)) yields immediately, because of (6.10),

$$\frac{d}{d(\lambda^2/2)} M\begin{pmatrix} \overset{\circ}{p}_1 \overset{\circ}{p}_2 \\ \overset{\circ}{q}_1 \overset{\circ}{q}_2 \end{pmatrix} = \int d\xi_1 \int d\xi_2 \, \Sigma \, \gamma^1 \gamma^2 [\xi_1 \xi_2] \, M\begin{pmatrix} \overset{\circ}{p}_1 \overset{\circ}{p}_2 \xi_1 \xi_2 \\ \overset{\circ}{q}_1 \overset{\circ}{q}_2 \xi_1 \xi_2 \end{pmatrix}. \tag{6.11}$$

The integration $\int d\xi_1 \int d\xi_2$ gives infrared divergences whenever ξ_1, ξ_2 are in positions such that

    (a) a pair p_1, p_2 or q_1, q_2 is directly connected with ξ_1, ξ_2;

    (b) a pair p, q is directly connected with ξ_1, ξ_2;

    (c) one momentum state p or q is connected to itself by ξ_1 or ξ_2 through an ultraviolet self-energy point.

Typical instances of these situations are shown in Fig. 1, to which all those resulting from the appropriate symmetrizations must be added.

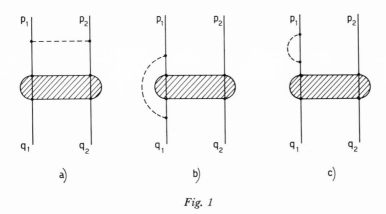

Fig. 1

Figure 1c shows the interference of ultraviolet with infrared divergences, which is assumed to be taken care of by some renormalization.

We neglect all other integrations that are contained in the perturbative expansion of (6.11) and decompose $\int d\xi_1 \int d\xi_2$ as follows:

$$\int d\xi_1' \int d\xi_2\, f(\xi_1, \xi_2, \ldots) = \hat{\int} d\xi_1 \hat{\int} d\xi_2\, f(\xi_1, \xi_2, \ldots) + \hat{D}^{12}(\xi_1, \xi_2, \ldots)$$

$$(6.12)$$

$\hat{\int} d\xi_1 \hat{\int} d\xi_2\, f$ in (6.12) will be such that it coincides with $\int d\xi_1 \int d\xi_2 f$ whenever the latter is convergent (and then $\hat{D}^{12} f \equiv 0$); it will stay finite if $\int d\xi_1 \int d\xi_2 f$ is infrared divergent, the divergent part being given by $\hat{D}^{12}$. Since we work with a renormalized theory, there are no other infinities to worry about.

$\hat{\int} d\xi_1 \hat{\int} d\xi_2$ is still far from being uniquely defined at this stage, nor do we wish to discuss here the optimum way to define it for practical purposes. We prefer to take advantage of the fact that our considerations apply as well to perturbative expansions, which permits us to use in (6.12) a definition that agrees entirely with the standard procedure.

With $\int d\xi_1 \int d\xi_2$ we always find in (6.11) the integrand factor

$$[\xi_1 \xi_2] = \frac{-i}{(2\pi)^4}\, \delta_{\mu_1\mu_2} \int d^4k\, \frac{1}{k^2 - i\epsilon}\, \exp[ik(\xi_1 - \xi_2)];$$

$$(6.13)$$

we define

$$\hat{\int} d\xi_1 \hat{\int} d\xi_2 \, \gamma^1\gamma^2[\xi_1\xi_2] g(\xi_1, \xi_2, \ldots) = \lim_{\rho \to 0} \int d\xi_1 \int d\xi_2 \, \gamma^1\gamma^2$$

$$\left\{ \frac{-i}{(2\pi)^4} \, \delta_{\mu_1\mu_2} \left[\int_{|\mathbf{k}|>K} d^4k \, \frac{1}{k^2 - i\epsilon} \right. \right.$$

$$\exp[ik(\xi_1 - \xi_2)] + \int_{|\mathbf{k}|<K} d^4k \, \frac{1}{k^2 - i\epsilon} \left(\exp[ik(\xi_1 - \xi_2)] \right.$$

$$\left. \left. \left. - \rho^2 \exp[i\rho k(\xi_1 - \xi_2)] \right) \right] \right\} g(\xi_1, \xi_2, \ldots), \tag{6.14}$$

and

$$\hat{D}^{12}\gamma^1\gamma^2[\xi_1\xi_2] g(\xi_1,\xi_2, \ldots)$$

$$= \lim_{\rho \to 0} \int d\xi_1 \int d\xi_2 \gamma^1\gamma^2 \left\{ \frac{-i}{(2\pi)^4} \, \delta_{\mu_1\mu_2}\rho^2 \int_{|\mathbf{k}|<K} d^4k \, \frac{1}{k^2 - i\epsilon} \right.$$

$$\left. \exp[i\rho k(\xi_1 - \xi_2)] \right\} g(\xi_1, \xi_2, \ldots) \tag{6.15}$$

where K is an arbitrary positive cutoff, which is taken to be sufficiently small in the next section.

Elementary computations prove that the definitions (6.14), (6.15) satisfy all the wanted requirements: the symbol $\hat{\int}$ differs from the old one $\int$ only when there are infrared divergences, which are set apart in (6.16). We find

$$\frac{d}{d(\lambda^2/2)} M\begin{pmatrix} \mathring{p}_1\mathring{p}_2 \\ \mathring{q}_1\mathring{q}_2 \end{pmatrix} = \hat{\int}\hat{\int} d\xi_1 \, d\xi_2 \, \gamma^1\gamma^2[\xi_1\xi_2] M\begin{pmatrix} \mathring{p}_1\mathring{p}_2 \, \xi_1\xi_2 \\ \mathring{q}_1\mathring{q}_2 \, \xi_1\xi_2 \end{pmatrix}$$

$$+ \hat{D}^{12}\gamma^1\gamma^2[\xi_1\xi_2] M\begin{pmatrix} \mathring{p}_1\mathring{p}_2 \, \xi_1\xi_2 \\ \mathring{q}_1\mathring{q}_2 \, \xi_1\xi_2 \end{pmatrix} \tag{6.16}$$

where the divergent term is computed explicitly from (6.9), (6.11), and (6.15), since the terms in the expansion of

$$\begin{pmatrix} \mathring{p}_1 \, \mathring{p}_2 \, \xi_1 \cdots \xi_{2n+2} \\ \mathring{q}_1 \, \mathring{q}_2 \, \xi_1 \cdots \xi_{2n+2} \end{pmatrix}$$

that give rise to infrared infinities such as (a), (b), (c) of Fig. 1, can be easily studied one by one.

A straightforward calculation then gives

$$\hat{D}^{12}\gamma^1\gamma^2[\xi_1\xi_2]M\begin{pmatrix}\mathring{p}_1\mathring{p}_2\,\xi_1\xi_2\\\mathring{q}_1\mathring{q}_2\,\xi_1\xi_2\end{pmatrix}=-IM\begin{pmatrix}\mathring{p}_1\mathring{p}_2\\\mathring{q}_1\mathring{q}_2\end{pmatrix}, \tag{6.17}$$

with

$$I \equiv J(p_1, p_2) + J(q_1, q_2) - J(p_1, q_1) - J(p_2, q_2) - J(p_1, q_2)$$
$$- J(p_2, q_1) + \tfrac{1}{2}[J(p_1, p_1) + J(p_2, p_2) + J(q_1, q_1) + J(q_2, q_2)], \tag{6.18}$$

where

$$J(p, q) = -\frac{2i}{(2\pi)^4}\,(p_1 \cdot q)\int_{|\mathbf{k}|<K} d^4k\,[k^2(p\cdot k)(q\cdot k)]^{-1}. \tag{6.19}$$

For example, in the case of the confluence of the pair ξ_1 and ξ_2, with p_1 and p_2, we have

$$\hat{D}^{12}(p_1 p_2)\gamma^1\gamma^2[\xi_1\xi_2]M\begin{pmatrix}\mathring{p}_1\mathring{p}_2\,\xi_1\xi_2\\\mathring{q}_1\mathring{q}_2\,\xi_1\xi_2\end{pmatrix}$$

$$= \frac{+i}{(2\pi)^4}\,\bar{u}(p_1)\bar{u}(p_2)\gamma_\mu{}^1\gamma_\mu{}^2\lim_{\rho\to 0}\rho^2$$

$$\times \lim_{\rho\to 0}\rho^2\int d^4k\,\frac{1}{k^2}\,\frac{i\gamma^1(p_1-\rho k)-m}{(p_1-\rho k)^2+m^2}\,\frac{i\gamma^2(p_2+\rho k)-m}{(p_2+\rho k)^2+m^2}$$

$$= -J(p_1, p_2)M\begin{pmatrix}\mathring{p}_1\mathring{p}_2\\\mathring{q}_1\mathring{q}_2\end{pmatrix} \tag{6.20}$$

after the limit is taken. I is therefore given by

$$I = -\frac{i}{(2\pi)^4}\int_{|\mathbf{k}|<K} d^4k\,\frac{1}{k^2-i\epsilon}\,Q^2(k), \tag{6.21}$$

with

$$Q(k) = \frac{p_1}{(p_1\cdot k)} + \frac{p_2}{p_2\cdot k} - \frac{q_1}{(q_1\cdot k)} - \frac{q_2}{(q_2\cdot k)}. \tag{6.22}$$

With exactly the same arguments we would find, more generally,

$$\frac{d}{d(\lambda^2/2)} M\begin{pmatrix}\overset{\circ}{p}_1\overset{\circ}{p}_2\, x_1\cdots x_r\\ \overset{\circ}{q}_1\overset{\circ}{q}_2\, y_1\cdots y_r\end{pmatrix} = \int d\xi_1 \int d\xi_2\, \gamma^1\gamma^2[\xi_1\xi_2]$$

$$M\begin{pmatrix}\overset{\circ}{p}_1\overset{\circ}{p}_2\, \xi_1\xi_2\, x_1\cdots x_r\\ \overset{\circ}{q}_1\overset{\circ}{q}_2\, \xi_1\xi_2\, y_1\cdots y_r\end{pmatrix}$$

$$-IM\begin{pmatrix}\overset{\circ}{p}_1\overset{\circ}{p}_2\, x_1\cdots x_r\\ \overset{\circ}{q}_1\overset{\circ}{q}_2\, y_1\cdots y_r\end{pmatrix}, \tag{6.23}$$

which is true also if $y_k \equiv x_k$.

We see therefore that changing all matrix elements in (6.23) by a multiplicative factor gives

$$M\begin{pmatrix}\overset{\circ}{p}_1\overset{\circ}{p}_2\\ \overset{\circ}{q}_1\overset{\circ}{q}_2\end{pmatrix} = \exp\left[-\frac{\lambda^2 I}{2}\right] M_c\begin{pmatrix}\overset{\circ}{p}_1\overset{\circ}{p}_2\\ \overset{\circ}{q}_1\overset{\circ}{q}_2\end{pmatrix}, \tag{6.24}$$

where I is given by (6.21), This proves that infrared divergences, computed with the prescription used here (which is the close counterpart of the one usually followed), can indeed be factored out in an exponential factor. This result is independent of perturbative methods.

B. Soft-Photon Emission

It follows from (6.24) that the cross section for the process is

$$\sigma = \exp(-\lambda^2 I)\sigma_c. \tag{6.25}$$

We must consider now the cross section for the emission of n photons with momenta $k_1, \ldots, k_n$, respectively. The corresponding matrix element is given by

$$M\begin{pmatrix}\overset{\circ}{p}_1\overset{\circ}{p}_2\\ \overset{\circ}{q}_1\overset{\circ}{q}_2\end{pmatrix}k_1\cdots k_n\Bigg) = \lambda^n \int d\xi_1\cdots \int d\xi_n\, \Sigma\, \gamma^1\cdots\gamma^n \bar{Z}_{k_1}(\xi_1)\cdots \bar{Z}_{k_n}(\xi_n)$$

$$M\begin{pmatrix}\overset{\circ}{p}_1\overset{\circ}{p}_2\, \xi_1\cdots \xi_n\\ \overset{\circ}{q}_1\overset{\circ}{q}_2\, \xi_1\cdots \xi_n\end{pmatrix} \tag{6.26}$$

($Z_{k_i}(x)$ is the wave function of the photon with momentum k.) We are interested in the limit for vanishing $k_1, \ldots, k_n$. Using the same terminology as before, divergences with respect to ξ_1 when $k_1 \to 0$ may appear in the cross section when ξ_1 has confluences with either of the external p_1, p_2, q_1, and q_2. We define, similarly,

$$\int d\xi_1 = \int d\xi_1 + \hat{D}^1, \tag{6.27}$$

and neglect the finite contributions from $\hat{\int} d\xi_1$. The remaining divergent part is easily calculated from (6.26) and (6.27), with considerations identical to those made previously, as

$$\frac{\lambda}{(2\pi)^{3/2}} \frac{1}{(2k_1{}^0)^{1/2}} \left[\frac{(p_1 \cdot \epsilon_1)}{(p_1 \cdot k_1)} + \frac{(p_2 \cdot \epsilon_1)}{(p_2 \cdot k_1)} - \frac{(q_1 \cdot \epsilon_1)}{(q_1 \cdot k_1)} - \frac{(q_2 \cdot \epsilon_1)}{(q_2 \cdot k_1)} \right]$$

$$M\left(\left. \begin{matrix} \overset{\circ}{p}_1 \overset{\circ}{p}_2 \\ \overset{\circ}{q}_1 \overset{\circ}{q}_2 \end{matrix} \right| k_2 \cdots k_n \right), \tag{6.28}$$

where ϵ_1 is the polarization vector of the photon k_1. By induction we find that the divergent part of (6.27) is given by

$$\frac{\lambda^n}{(2\pi)^{3n/2}} \prod_{j=1}^{n} \frac{1}{(2k_j{}^0)^{1/2}} \left[\frac{(p_1 \cdot \epsilon_j)}{(p_1 \cdot k_j)} + \frac{(p_2 \cdot \epsilon_j)}{(p_2 \cdot k_j)} - \frac{(q_1 \cdot \epsilon_j)}{(q_1 \cdot k_j)} - \frac{(q_2 \cdot \epsilon_j)}{(q_2 \cdot k_j)} \right]$$

$$M\left(\begin{matrix} \overset{\circ}{p}_1 \overset{\circ}{p}_2 \\ \overset{\circ}{q}_1 \overset{\circ}{q}_2 \end{matrix} \right), \tag{6.29}$$

so that the cross section for n soft-photon emission has a divergent part given by

$$\sigma_n = \frac{f^n}{n!} \exp[-\lambda^2 \cdot I] \sigma_c \tag{6.30}$$

where

$$f = \frac{\lambda^2}{(2\pi)^3} \sum_{\text{pol}} \int_{|k| < K} d^3k \, \frac{1}{2k_0} (Q \cdot \epsilon)^2, \tag{6.31}$$

with Q given by Eq. (6.22). It is then easy to see that

$$f = \lambda^2 I \tag{6.32}$$

because, since $Q \cdot k = 0$, $\sum_{\text{pol}} (Q \cdot \epsilon)^2 = Q^2$. Equation (6.30) thus reduces to

$$\sigma_n = \frac{f^n}{n!} \exp[-f] \sigma_c, \tag{6.33}$$

which is the desired Poisson distribution originally given by Block and Nordsieck. The total cross section is obviously

$$\sigma_{\text{tot}} = \sum_{n=0}^{\infty} \sigma_n = \sigma_c, \tag{6.34}$$

which exhibits the cancellation of the infrared divergences against the divergent contributions due to soft-photon emission.

Energy conservation is neglected in Eq. (6.34). The condition $k \to 0$ is unrealistic; if we impose, instead, for all n,

$$\sum_{i=1}^{n} k_i{}^0 < \Delta E, \tag{6.35}$$

we have the standard result

$$\sigma_{\text{tot}} \left(\Delta E \right) \simeq F(\beta) \, \exp\left[\beta \, \log \frac{\Delta E}{K} \right] \sigma_c \tag{6.36}$$

where

$$\beta = \frac{\lambda^2}{2(2\pi)^3} \int d\Omega_k \, |k|^2 \, Q^2, \tag{6.37}$$

and

$$F(\beta) = \frac{1}{2\pi i} \int_{-\infty}^{+\infty} dt \, \frac{\exp[it]}{t - i\epsilon} \exp\left[\beta \int_0^1 dx \, \frac{\exp[-itx] - 1}{x} \right]$$

$$= 1 - \frac{\pi^2}{12} \beta^2 + 0(\beta^4), \tag{6.38}$$

so that the divergent term

$$f = \beta \int_0^k dk \, \frac{1}{k} \tag{6.39}$$

disappears.

Part III

Regularization, Renormalization, and Mass Equations

Chapter 7

x-Space Regularization

1. REGULARIZATION ON PAIRS OF POINTS

This chapter deals with x-space regularization in general, without reference to renormalization, which will be the subject of the following chapters. This sequence is convenient because x-space regularization, as defined here, can also be applied quite generally to theories that are not renormalizable; or we may only wish to know how to compute a single graph in x space, in a manner that yields the correctly renormalized expression when the theory is renormalizable.

A major difference between x-space and p-space regularization is that in the first case we *can* work with *pairs of points*; in the other, we *have* to work with the *lines* joining those points, which may be many more than the pairs. Our treatment will be based upon distribution theory, whence, for convenience, we shall derive the notion of "finite-part integral" in Section 4.

A typical connected graph $\mathcal{G}(\xi, \mathcal{L})$, such as may occur in any perturbative expansion, is defined as a set $\mathcal{L} \equiv \{l_1, l_2, \ldots, l_L\}$ of *lines* joining the *vertices* $\xi \equiv \{\xi_1, \xi_2, \ldots, \xi_n\}$; if ξ_{i_l} and ξ_{j_l} are, respectively, the incoming and outgoing vertices of the line l, and $D_{F,l}$ is the corresponding propagator for that line, $\mathcal{G}(\xi, \mathcal{L})$ yields the product of distributions

$$T(\xi) = \prod_{l \in \mathcal{L}} D_{F,l}(\xi_{i_l} - \xi_{j_l}; m_l). \qquad (7.1)$$

Expressions such as $T(\xi)$ are meaningless and cannot be considered as distributions in $S'(R^{4n})$ [44]. However, it is well known that we can proceed as follows [45].

(a) Define a subspace $S_0(R^{4n})$ of $S(R^{4n})$ containing only test functions $\phi(\xi_1, \ldots, \xi_n)$ which vanish with zeros of suitably high orders when any two variables are coincident; then (7.1) defines a continuous linear functional.

(b) Extend $T(\xi)$ from $S_0'(R^{4n})$ to the whole $S'(R^{4n})$: this extension is *not* unique, and therein lies the typical ambiguity of any regularization procedure (which must be proved to be physically irrelevant–renormalizab. theories; or suppressed with *ad hoc* rules–nonrenormalizable theories).

Regularization can be performed in many ways, as is well illustrated by the literature. We shall limit ourselves to the discussion of two procedures based upon analytic continuation [46–50], both well suited to our purposes; they are denoted for short the σ and $\hat{\delta}$ procedures. The first is discussed here, the second in the following section.

Introduce the auxiliary distributions [49]

$$\sigma(x, \lambda) = [\mu^2(x^2 - i0^+)]^\lambda \tag{7.2}$$

$(x^2 = x_0^2 - \mathbf{x}^2; \sigma$ is a meromorphic distribution with poles of first order in $\lambda = -2, -3, \ldots)$, and define, from (7.1),

$$T(\xi, \lambda) = T(\xi)\sigma(\xi_1 - \xi_2, \lambda_{12}) \cdots \sigma(\xi_{n-1} - \xi_n, \lambda_{n-1,n}) \tag{7.3}$$

with $i < j$, λ_{ij} complex; $\lambda \equiv \{\lambda_{12}, \lambda_{13}, \ldots, \lambda_{n-1,n}\}$. Graphically,

(broken lines denote the auxiliary distributions $\sigma(\xi_i - \xi_j, \lambda_{ij})$).

It can then be proved [49] that $T(\xi, \lambda)$ defines a distribution that is holomorphic in some region Ω of C, where C is the tensor product of the complex planes of all the parameters λ. Furthermore, $T(\xi, \lambda)$ can be continued to a distribution $\widetilde{T}(\xi, \lambda)$ which is meromorphic in the whole C. Finally, we can define, starting from $\widetilde{T}(\xi, \lambda)$, the desired extension of $T(\xi)$ (acting in $S.$ (R^{4n})) to a $\widetilde{T}(\xi)$ which acts on $S(R^{4n})$, by means of the operation $\theta(\lambda)$, which is defined later in (7.5).

Denote with $\theta(\lambda)\phi(\lambda)$ the limit for $\lambda \to 0$ of the *regular part* of the Laurent expansion of the meromorphic function $\phi(\lambda)$ around $\lambda = 0$.

Also, define $\theta(\lambda_i)$ as acting in the same manner on the variable λ_i of $\phi(\lambda_1, \lambda_2, \ldots, \lambda_n)$, all other variables being kept fixed.

Define next

$$\theta(\lambda_h \lambda_k) = \frac{1}{2!} [\theta(\lambda_h)\theta(\lambda_k) + \theta(\lambda_k)\theta(\lambda_h)] = S[\theta(\lambda_h)\theta(\lambda_k)]$$

and $\theta(\lambda_1 \lambda_2 \cdots \lambda_j)$ by symmetrization of the operations $\theta(\lambda_h)$ in all $j!$ ways. Assign an arbitrary function $f(\lambda)$ of λ holomorphic at $\lambda = 0$, with $f(0) = 1$. Then the wanted extension of $T(\xi)$ is

$$\tilde{T}(\xi) = \bar{\theta}(\lambda)f(\lambda_{12}) \cdots f(\lambda_{n-1,n})\tilde{T}(\xi, \lambda) \tag{7.4}$$

with (Σ_P is over all permutations $i_1, \ldots, i_n$)

$$\bar{\theta}(\lambda) = \frac{1}{n!} \sum_P [\theta(\lambda_{i_1, i_2})\theta(\lambda_{i_1, i_3}\lambda_{i_2, i_3}) \cdots \theta(\lambda_{i_1 i_n} \cdots \lambda_{i_{n-1}, i_n})]$$

$$\equiv S[\theta(\lambda_{1,2})\theta(\lambda_{1,3}\lambda_{2,3}) \cdots \theta(\lambda_{i,n} \cdots \lambda_{n-1,n})]. \tag{7.5}$$

The proof, reported in [49], confirms that only the number of vertices, not that of lines, is relevant. We report here only some essential traits of it. We remark first that

$$D_{F,l}(x; m_l) = d_l(\partial) \Delta_F(x, m_l)$$

where $d_l(\partial)$ is some differential operator (in a distribution-theoretical sense) whose nature depends on that of the particle, and $\Delta_F(x, m_l)$ is the free propagator corresponding to mass m_l and spin 0:

$$\Delta_F(x, m) = \frac{1}{4i\pi^2} \frac{1}{x^2 - i0^+} + \frac{im^2}{16\pi^2} \log[m^2(x^2 - i0^+)]D_1(x, m)$$

$$+ \frac{im^2}{16\pi^2} D_2(x, m). \tag{7.6}$$

We shall restrict our treatment, without loss of generality, to the case in which $D_{F,l}(x, m_l)$ is replaced simply with $\Delta_F(x, m_l)$.

It appears from (7.6) that the only singular terms to be handled, as a possible cause of ultraviolet divergences, are of type

$$p_{hk}(x) = (x^2 - i0^+)^{-h} \log^k(x^2 - i0^+), \tag{7.7}$$

everything else giving only multipliers in $S(R^{4n})$. Furthermore, suitable derivatives with respect to λ in $\sigma(x, \lambda)$ reproduce any wanted power k

of $\log(x^2 - i0^+)$: this shows that, in conclusion, we may consider, in the place of (7.3), more simply a product

$$\sigma(\xi_1 - \xi_2, \lambda_{12})\sigma(\xi_1 - \xi_3, \lambda_{13}) \cdots \sigma(\xi_{n-1} - \xi_n, \lambda_{n-1, n}) \qquad (7.8)$$

and then take $\lambda_{ij} = -h_{ij}$ after a suitable number of differentiations. The proof [49] goes through the following steps, which we only mention here.

(a) take (ϵ, the same for all σ_ϵ, $\rightarrow 0$)

$$\sigma_\epsilon(x, \lambda) \equiv (x^2 - i\epsilon)^\lambda = \frac{\exp[-i\lambda\pi/2]}{\Gamma(-\lambda)} \int_0^\infty \exp\left[-i\,\frac{x^2 - i\epsilon}{\alpha}\right]\alpha^{\lambda+1}\,\frac{d\alpha}{\alpha^2}$$

($\mathrm{Re}\,\lambda < 0; \mu \equiv 1$).

(b) Take the Fourier transform of (7.8), introducing for each vertex its conjugate p_i.

(c) Prove that the expression thus obtained in the variables p_i is analytic in a region Ω of $C(\lambda$ plane) and can be continued to a meromorphic distribution.

The splitting and separate consideration made earlier, for simplicity's sake, of the terms $\Delta_F(x, m)$ in (7.6) might seem to imply the risk of unwanted infrared divergences. The reader can easily see, though, that this cannot be the case for the complete expression.

2. REGULARIZATION BY SPLITTING OF COINCIDENT POINTS

In the previous section we have introduced the σ distribution in order to give meaning to the expression (7.1); in the present section we show another method [51, 52] which enables us to achieve an equivalent result. This approach is particularly convenient in the study of point-loop ambiguities or tadpoles. Let us introduce the distribution

$$\hat{\delta}(x, \lambda) = \frac{i}{\pi^2}\,\lambda f(\lambda)(x^2 - i\epsilon)^{\lambda - 2} \qquad (7.9)$$

where $f(\lambda)$ is an arbitrary function of λ holomorphic in the neighborhood of $\lambda = 0$, $f(0) = 1$. $\hat{\delta}$ is a meromorphic distribution, with singularities that are poles of first order in $\lambda = 0, -1, -2, \ldots$. Clearly

$$\lim_{\lambda \to 0} \hat{\delta}(x, \lambda) = \delta(x).$$

We define next the quantities

$$D_{F,l}^{\lambda,\mu}(x, y; m) = d_l(\partial)\, \Delta_F^{\lambda,\mu}(x, y; m)$$

$$= d_l(\partial) \int d\eta_1 \int d\eta_2\, \hat{\delta}(\lambda, x - \eta_1)\, \hat{\delta}(\mu, y - \eta_2)$$

$$\times \Delta_F(\eta_1 - \eta_2; m)$$

$$= \frac{\lambda\mu \exp[-i\,(\pi/2)(\lambda + \mu)]\, f(\lambda)\, f(\mu)}{\Gamma(1 - \lambda)\Gamma(1 - \mu)} \int_0^1 dr_1 \int_0^1 dr_2$$

$$\times (1 - r_1)^{\lambda - 1}(1 - r_2)^{\mu - 1} d_l(\partial)\, \Delta^{\lambda + \mu - 1}(x - y; r_1 r_2 m)$$

where

$$\Delta^\lambda(x, m) = \frac{1}{16\pi^2} \int_0^\infty \frac{ds}{s^2}\, s^\lambda \exp\left(-i\,\frac{x^2 - i\epsilon}{4s}\right) \exp\left(-im^2 s\right).$$

Now replace (7.1) with the distribution

$$T_1(\xi, \lambda) = \prod_{i < j}^n \prod_{r=1}^{K_j} D_F^{\lambda_i^{(s_r^{ij})} \lambda_j^{(s_r^{ji})}}(\xi_i - \xi_j; m).$$

where all parameters $\lambda_i^{(jr)}$ are associated to the point x_i, and the label $s_r^{(ij)}$ numbers the K_{ij} lines joining x_i with x_j; the total number K_i of labels $s_r^{(ij)}$ tells, for each given i, how many lines are confluent into x_i. The meaning of this replacement is made clear by remarking that the graph corresponding to $T_1(\xi, \lambda)$ is obtained from that of $T(\xi)$ through the splitting of confluent points.

$$T(\xi_1 \xi_2) = \quad \Rightarrow T_1(\xi_1 \xi_2, \lambda_1 \lambda_2) \quad = \quad ,$$

$$T(\xi_1 \cdots \xi_4) = \quad \Rightarrow T_1(\xi_1 \cdots \xi_4, \lambda_1 \cdots \lambda_4) = \quad ,$$

$$T(\xi_1 \cdots \xi_5) = \quad \Rightarrow T(\xi_1 \cdots \xi_5, \lambda_1 \cdots \lambda_5) =$$

(wavy lines denote the auxiliary distributions $\hat{\delta}$). The distribution $T_1(\xi, \lambda)$ can, in a trivial way, be connected with the distribution

$$\prod_{l \in \mathscr{L}} d_l(\partial) \Delta_F^{\lambda_l}(\xi_{i_l} - \xi_{j_l}; m_l)$$

studied by E. R. Speer [50]. His work suffices to show that the operation

$$\hat{\theta}(\lambda) = S[\theta(\lambda_1^{(1)} \cdots \lambda_1^{(K_1)}) \cdots \theta(\lambda_n^{(1)} \cdots \lambda_n^{(K_n)})]$$

with

$$\theta(\lambda_i^{(1)} \cdots \lambda_i^{(K_i)}) = S[\theta(\lambda_i^{(1)}) \cdots \theta(\lambda_i^{(4)})] \tag{7.10}$$

yields an extension of $T(\xi)$ to a $\hat{T}(\xi)$ that acts on $S(R^{4n})$:

$$\hat{T}(\xi) = \hat{\theta}(\lambda)\hat{T}(\xi, \lambda)$$

where $\hat{T}(\xi, \lambda)$ denotes the analytic continuation of $T_1(\ , \lambda)$ in the complex plane of the parameters λ.

3. EVALUATION OF "DIVERGENT" CONTRIBUTIONS [51]

Looking at (7.6), we note that

$$(\square - m^2)D_1(x, m) = 0, \tag{7.11a}$$

$$D_1(0, m) = 1, \tag{7.11b}$$

$$D_2(0, m) = 2c - 2\log 2 - 1 + i\pi, \tag{7.11c}$$

$$c = \lim_{K \to \infty} (c_k - \log(k+1)), \qquad c_k = \sum_{n=1}^{k} \frac{1}{n}; \tag{7.11d}$$

$$D_1(x, m) = \sum_{k=1}^{\infty} \frac{1}{k!(k-1)!} \left(-\frac{m^2 x^2}{2}\right)^{k-1}, \tag{7.12a}$$

$$D_2(x, m) = (2c - 2\log 2 + i\pi)D_1(x, m) - \sum_{k=1}^{\infty} \frac{2kc_k - 1}{(k!)^2} \left(-\frac{m^2 x^2}{4}\right)^{k-1} \tag{7.12b}$$

From (7.11) and (7.12) it is possible to compute the Laurent developments

$$(x^2 - i0^+)^{-h+\lambda} = \sum_{s=-1}^{\infty} T_s^{(h)}(x)\lambda^s, \qquad |\lambda| < 1, h = 2, 3, \ldots \tag{7.13}$$

We find

$$T_{-1}^{(h)}(x) = \frac{i\pi^2 (-1)^{h-1}}{4^{h-2}(h-2)!(h-1)!} \Box^{h-2} \delta^{(4)}(x),$$

$$T_0^{(h)}(x) = \frac{(-1)^h}{4^{h-1}[(h-1)!]^2} \Box^{h-1} \frac{1}{x^2 - i0^+}$$

$$\times [2(h-1)c_{h-1} - 1 + (h-1)\log(x^2 - i0^+)], \quad \text{etc.}$$

$$(7.14)$$

These results can be of great importance in the computation of divergent contributions. For instance, from (7.9), (7.13), and (7.14) we find

$$\theta(\lambda)\hat{\delta}(x, \lambda)(x^2 - i0^+)^{-1} = \tfrac{1}{8} \Box \, \delta(x) \stackrel{\text{def}}{=} \text{FP}(x^2 - i0^+)\delta(x),$$

where we have introduced the concept of *partie finie* (FP) of a product of distributions (we shall return to this in Chapter 9). More generally, from (7.7) and (7.14):

$$\text{FP } p_{hk}(x)\delta(x) = \frac{i}{\pi^2} \, \theta(\lambda)\lambda f(\lambda)(x^2 - i0^+)^{\lambda-2-h}\log^k(x^2 - i0^+)$$

$$= \frac{i}{\pi^2} \, T_{-1}^{(h+2)}(-1)^k k! f_{(0)}^{(k)}. \qquad (7.15)$$

Of special interest to us is the case where $h = 0$, for which (7.15) yields

$$\text{FP} \log(x^2 - i0^+)\delta(x) = -f^{(1)}(0)\delta(x), \qquad (7.16)$$

so that the point loop

is given, after a trivial computation, by

$$\text{FP } \Delta_F(0, m) = \frac{m^2}{16\pi^2} \log \frac{m^2}{\mu^2} \qquad (7.17)$$

if we set

$$f^{(1)}(0) = i\pi^2 + 2c - 2\log 2 - 1 + \log \mu^2.$$

Use of these formulas greatly facilitates all calculations that occur in renormalization theory.

4. FINITE-PART INTEGRALS AS SHORTHAND

We want to show that regularizations of the nature discussed thus far can be formally indicated by means of *finite-part integrals* $\int$ [15, 16, 49]; all formal work in the manipulation of equations and expansions in renormalizable theories will be greatly simplified by use of this convention.

We introduce first the concept of integral of a distribution. Let $T(\xi_1 \cdots \xi_n)$ be a distribution defined in a functional space $\phi(\xi_1 \cdots \xi_n)$:

$$T(\xi_1 \cdots \xi_n) : \Phi(\xi_1 \cdots \xi_n) \to \mathscr{C}, \qquad \phi(\xi_1 \cdots \xi_n) \to$$

$$\to \langle T(\xi_1 \cdots \xi_n), \phi(\xi_1 \cdots \xi_n) \rangle,$$

$$\langle T(\xi_1 \cdots \xi_n), \phi(\xi_1 \cdots \xi_n) \rangle \equiv \int \cdots \int d\xi_1 \cdots d\xi_n$$

$$\times T(\xi_1 \cdots \xi_n)\phi(\xi_1 \cdots \xi_n), \tag{7-18}$$

where the integrations have only a symbolic meaning, and are to be considered only as a device to denote that the variables $\xi_1 \cdots \xi_n$ are completely "saturated" in $\langle , \rangle$.

Define next $\int d\xi_n T(\xi_1 \cdots \xi_n)\phi(\xi_1 \cdots \xi_n)$ as a functional in a space $\Phi(\xi_1 \cdots \xi_{n-1})$ defined by

$$\langle \int d\xi_n T(\xi_1 \cdots \xi_n)\phi(\xi_1 \cdots \xi_n), \psi(\xi_1 \cdots \xi_{n-1}) \rangle$$

$$= \langle T(\xi_1 \cdots \xi_n), \phi(\xi_1 \cdots \xi_n)\psi(\xi_1 \cdots \xi_{n-1}) \rangle; \tag{7.19}$$

we note that the foregoing functional is well defined also when $\psi = 1$. We can likewise introduce successive integrations, and it is immediately seen that the final result (i.e., all n variables saturated) does not depend on the order in which the "integrations" are performed.

In order to introduce the concept of finite-part integral of distributions [49] we consider the classes $\mathscr{F}^{(n)}$ of formal distributions $T(\xi_1 \cdots \xi_n)$ of the type we have been considering in the previous sections. Let us choose a specific $T(\xi_1 \cdots \xi_n)$; it will in general be defined in a certain space Φ_0 (which may in particular coincide with the whole $S(R^{4n})$) whose functions can be subjected to the condition that they vanish with zeros of suitably high order when two variables coincide. Denote now with $T_{n,n-1}(\xi_1 \cdots \xi_n)$ the extension of $T(\xi_1 \cdots \xi_n)$, obtained with some specified rule, into a space of functions on whose behavior we may impose conditions when any two variables happen to be confluent but whose behavior when $\xi_n \to \xi_{n-1}$ is in no way restricted. A concrete realization of $T_{n,n-1}$ is the one proved previously:

$$T_{n,n-1}(\xi_1 \cdots \xi_n) = \theta(\lambda)\sigma(\xi_n - \xi_{n-1}; \lambda)T(\xi_1 \cdots \xi_n). \tag{7.20}$$

We denote by

$$\underset{\xi_{n-1}}{\int} d\xi_n \, T(\xi_1 \cdots \xi_n)\phi(\xi_1 \cdots \xi_n) \tag{7.21}$$

a functional that is defined in function space $\Phi(\xi_1 \cdots \xi_{n-1})$ as

$$\langle \underset{\xi_{n-1}}{\int} d\xi_n \, T(\xi_1 \cdots \xi_n)\phi(\xi_1 \cdots \xi_n), \, \psi(\xi_1 \cdots \xi_{n-1}) \rangle$$

$$= \langle T_{n,n-1}(\xi_1 \cdots \xi_n), \, \phi(\xi_1 \cdots \xi_n)\psi(\xi_1 \cdots \xi_{n-1}) \rangle,$$

we note that we may take $\psi = 1$ provided ϕ is suitably restricted, except for $\xi_n \to \xi_{n-1}$.

We can likewise regularize also the confluence $\xi_n \to \xi_1 \cdots \xi_{n-2}$ by introducing the distribution $T_n(\xi_1 \cdots \xi_n)$, which is defined in a suitable functional space upon whose elements no restriction is imposed concerning their behavior with respect to the variable ξ_n; for example,

$$T_n(\xi_1 \cdots \xi_n) = S[\theta(\lambda_{n1}) \cdots \theta(\lambda_{n,n-1})] \, T(\xi_1 \cdots \xi_n)\sigma(\xi_n - \xi_1; \lambda_{n1})$$

$$\cdots \sigma(\xi_n - \xi_{n-1}; \lambda_{n,n-1}).$$

Define next

$$\underset{\xi_1 \cdots \xi_{n-1}}{\int} d\xi_n \, T(\xi_1 \cdots \xi_n)\phi(\xi_1 \cdots \xi_n) \tag{7.22}$$

by means of

$$\langle \underset{\xi_1 \cdots \xi_{n-1}}{\int} d\xi_n \, T(\xi_1 \cdots \xi_n)\phi(\xi_1 \cdots \xi_n), \, \psi(\xi_1 \cdots \xi_{n-1}) \rangle$$

$$= \langle T_n(\xi_1 \cdots \xi_n), \, \phi(\xi_1 \cdots \xi_n)\psi(\xi_1 \cdots \xi_{n-1}) \rangle.$$

If ϕ is not restricted in any way, the foregoing functional is defined in general only for particular functions ψ having suitable zeros. Let us suppose that the chosen extension rule is such that the functional obtained in this way can be extended, by means of the same rule, to functions on which no restriction is imposed as regards the behavior of ψ with respect to ξ_{n-1}.

We thus arrive at the definition

$$\underset{\xi_1 \cdots \xi_{n-1}}{\int} d\xi_{n-1} \psi(\xi_1 \cdots \xi_{n-1}) \underset{\xi_1 \cdots \xi_{n-1}}{\int} d\xi_n \, T(\xi_1 \cdots \xi_n)\phi(\xi_1 \cdots \xi_n)$$

as functional in a $\phi(\xi_1 \cdots \xi_{n-2})$ and finally, continuing the iteration procedure, at

$$\int d\xi_1 \phi(\xi_1) \int d\xi_2 \phi(\xi_1 \xi_2) \cdots \int_{\xi_1 \cdots \xi_{n-1}} d\xi_n \, T(\xi_1 \cdots \xi_n) \phi(\xi_1 \cdots \xi_n)$$

$$(7.23)$$

as a complex number.

We shall prove now easily that by means of the continuation procedure discussed in the previous sections it is possible to give a well-defined meaning to the chain of integrals (7.23) where $\phi(\xi_1 \cdots \xi_n) \in S(R^{4n})$ and $\phi(\xi_1) = \cdots = \phi(\xi_1 \cdots \xi_{n-1}) = 1$, as specified.

We shall restrict ourselves to the detail concerning the case $n = 3$; after this analysis the general validity of our assertion will be clear.

From the functional $T(\xi_1 \xi_2 \xi_3)$ we can obtain, by extension, the functional $T_3(\xi_1 \xi_2 \xi_3)$, defined as

$$T_3(\xi_1 \xi_2 \xi_3) = S[0(\lambda_{32})0(\lambda_{31})] \, T(\xi_1 \xi_2 \xi_3) \sigma(\xi_3 - \xi_1; \lambda_{31}) \sigma(\xi_3 - \xi_2; \lambda_{32}),$$

which bears no restriction on the behavior of the test function as regards the variable ξ_3 (any restriction will therefore affect only the behavior when $\xi_1 \to \xi_2$).

Define now the functional $T(\xi_1 \xi_2) = \int_{\xi_1 \xi_2} d\xi_3 \, T(\xi_1 \xi_2 \xi_3) \phi(\xi_1 \xi_2 \xi_3)$

without restrictions on the $\phi(\xi_1 \xi_2 \xi_3)$ by requiring that it associate to the test function $\psi_0(\xi_1 \xi_2)$, which vanishes with zeros of suitable order when $\xi_1 \to \xi_2$, the number

$$\langle T(\xi_1 \xi_2), \psi_0(\xi_1 \xi_2) \rangle = \langle T_3(\xi_1 \xi_2 \xi_3), \phi(\xi_1 \xi_2 \xi_3) \psi_0(\xi_1 \xi_2) \rangle.$$

We obtain by extension from $T(\xi_1 \xi_2)$ the functional T_2:

$$\langle T_2(\xi_1 \xi_2), \psi(\xi_1 \xi_2) \rangle = \theta(\lambda_{12}) \langle T(\xi_1 \xi_2) \sigma(\xi_1 - \xi_2, \lambda_{12}), \psi(\xi_1 \xi_2) \rangle$$

$$= \theta(\lambda_{12}) \langle T_3(\xi_1 \xi_2 \xi_3) \sigma(\xi_1 - \xi_2; \lambda_{12}),$$

$$\phi(\xi_1 \xi_2 \xi_3) \psi(\xi_1 \xi_2) \rangle$$

$$= \theta(\lambda_{12}) S[\theta(\lambda_{13}) \theta(\lambda_{23})] \langle T(\xi_1 \xi_2 \xi_3)$$

$$\sigma(\xi_1 - \xi_2; \lambda_{12}) \sigma(\xi_1 - \xi_3; \lambda_{13}) \sigma(\xi_2 - \xi_3; \lambda_{23}),$$

$$\phi(\xi_1 \xi_2 \xi_3) \psi(\xi_1 \xi_2) \rangle.$$

In conclusion, by setting $\psi = 1$, we arrive at the definition

$$\int d\xi_1 \int_{\xi_1} d\xi_2 \int_{\xi_1 \xi_2} d\xi_3 \, T(\xi_1 \xi_2 \xi_3) = \theta(\lambda_{12}) S[\theta(\lambda_{13})\theta(\lambda_{23})]$$

$$\times \langle T(\xi_1 \xi_2 \xi_3)\sigma(\xi_1 - \xi_2; \lambda_{12})$$

$$\times \sigma(\xi_1 - \xi_3; \lambda_{13})\sigma(\xi_2 - \xi_3; \lambda_{23}),$$

$$\phi(\xi_1 \xi_2 \xi_3) \rangle;$$

it is then immediately seen, from the way in which we have defined the chain of integrations, that in general it is not possible to exchange arbitrarily the order of the integrations without causing alterations in the final results.

Thus, we may conclude by stating that the multiple integral

$$\int_{(\xi_1 \cdots \xi_n)} d\xi_1 \cdots d\xi_n \, T(\xi_1 \cdots \xi_n)\phi(\xi_1 \cdots \xi_n) \equiv \int d\xi_1 \int_{\xi_1} d\xi_2 \cdots$$

$$\times \int_{\xi_1 \cdots \xi_{n-1}} d\xi_n \, T(\xi_1 \cdots \xi_n)\phi(\xi_1 \cdots \xi_n) \tag{7.24}$$

can be defined by

$$\theta(\lambda_{12}) S[\theta(\lambda_{13})\theta(\lambda_{23})] \cdots S[\theta(\lambda_{1n}) \cdots \theta(\lambda_{n-1, n})]$$

$$\times \lim_{\epsilon \to 0^+} \int \cdots \int d\xi_1 \cdots d\xi_n \, T_\epsilon(\xi_1 \cdots \xi_n)\sigma(\xi_1 - \xi_2; \lambda_{12}) \cdots$$

$$\times \sigma(\xi_n - \xi_{n-1}; \lambda_{n, n-1})\phi(\xi_1 \cdots \xi_n)$$

but the order of partial integrations at the right-hand side in general cannot by freely exchanged.

It is, however, sufficient to define

$$\int_{\xi_1 \cdots \xi_{n-1}} d\xi_n \, T(\xi_1 \cdots \xi_n)\psi(\xi_1 \cdots \xi_n)$$

$$= \frac{1}{n} \int_{\xi_1 \cdots \xi_{n-1}} d\xi_n \, \{T(\xi_1 \cdots \xi_n)\psi(\xi_1 \cdots \xi_n) + (\xi_n \leftrightarrow \xi_1) + \cdots$$

$$+ (\xi_n \leftrightarrow \xi_{n-1})\}, \tag{7.25}$$

and so on, where $(\xi_n \leftrightarrow \xi_j)$ means

$$T(\xi_1 \cdots \xi_{j-1} \xi_n \xi_{j+1} \cdots \xi_j) \, \psi(\xi_1 \cdots \xi_{j-1} \xi_n \xi_{j+1} \cdots \xi_j),$$

to obtain complete symmetry.

Finally we shall note that by using the $\hat{\delta}$ procedure we can introduce the extension

$$T_n(\xi_1 \cdots \xi_n) = \left[\prod_i^k \theta(\lambda_n^{(i)}) \hat{\delta}(\xi_n^{(i)} - \xi_n, \lambda_n^{(i)}) \right]$$
$$\times T(\xi_1 \cdots \xi_{n-1} \xi_n^{(1)} \cdots \xi_n^{(i)} \cdots \xi_n^{(k)}) \qquad (7.26)$$

where $T(\xi_1 \cdots \xi_{n-1} \xi_n^{(1)} \cdots \xi_n^{(i)} \cdots \xi_n^{(k)})$ means that the nth variable was split in $\xi_n^{(1)} \cdots \xi_n^{(i)} \cdots \xi_n^{(k)}*$; and define the chain of integrals following the same procedure adopted before. The result is

$$\int d\xi_1 \int_{\xi_1} d\xi_2 \cdots \int_{\xi_1 \cdots \xi_{n-1}} d\xi_n \, T(\xi_1 \cdots \xi_n) \phi(\xi_1 \cdots \xi_n)$$
$$= \hat{\theta}(\lambda) \int dR_n \, \Gamma_1^{i_1} \cdots \Gamma_n^{i_n} T(\xi_1^{(1)} \cdots \xi_1^{(i_1)} \cdots \xi_n^{(1)} \cdots \xi_n^{(i_n)})$$
$$(7.27)$$

where $\hat{\theta}(\lambda)$ is given from (7.10) and

$$\Gamma_k^{i_k} = \prod_i^{i_k} \hat{\delta}(\xi_k^{(i)} - \xi_k; \lambda_k^{(i)})$$

$$\int dR_n = \int d\xi_1 \int d\xi_1^{(1)} \cdots \int d\xi_1^{(i_1)} \cdots \int d\xi_n \int d\xi_n^{(1)} \cdots \int d\xi_n^{(i_n)}.$$
$$(7.28)$$

Applications of these procedures are frequently made in the following chapters.

* For example, for

$$T(\xi_1 \xi_2 \xi_3) = D_{f,1}(\xi_1 - \xi_2) D_{F,1}(\xi_1 - \xi_3) D_{F,1}(\xi_2 - \xi_3)$$

we have

$$T_3(\xi_1 \xi_2 \xi_3^{(1)} \xi_3^{(2)}) = D_{F,1}(\xi_1 - \xi_2) D_{F,1}(\xi_1 - \xi_3^{(1)}) D_{F,1}(\xi_2 - \xi_3^{(2)}).$$

Chapter 8

A Superrenormalizable Theory

1. THE ϕ^3 THEORY

As an introduction to the general treatment of renormalizable
theories, where several formal complications may render less perspicuous
the process whereby the equations for propagators and Green's functions
are rendered well defined and physically meaningful, it is useful to
start our study of renormalization with the simpler case of a superrenormalizable theory, defined by the interaction Lagrangian
density [17]

$$\mathscr{L} = g\phi^3(x) + g_1\phi(x) \tag{8.1}$$

(neutral scalar field). This study [51] will show the following.

(i)   The $\hat{\delta}$ procedure defines the finite part of products of distributions in a way that makes all branching equations well defined and
"renormalized" because a change in the procedure will produce a new
renormalization, without changing the Green functions: only the
"bare" masses and changes are given new values.

(ii)   Furthermore, the perturbative expansions obtained from the
renormalized branching equations are *the same* as would be obtained by
expanding first and then renormalizing the expansions (with the same
prescription).

The same results will hold also for all renormalizable theories. All terms
subtracted or, if one prefers, arbitraries introduced by the regularization procedure will be seen to be point-support distributions, absorbable
into the bare parameters of the theory without damage to unitarity and
locality (as is of course wanted).

Our renormalization is *the same* as any other: working with the
branching equations simplifies combinatoric proofs, and may remain
significant even if, as is expected, the renormalized expansions will
diverge.

2. RENORMALIZATION OF BRANCHING EQUATIONS

We limit our discussion to type I equations for Green's functions:

$$G_n(x_1 \cdots x_n) = \sum_{j=2}^{n} Z[x_1 x_j] G_{n-2}(x_2 \cdots x_{j-1} x_{j+1} \cdots x_n)$$

$$+ 3ig \int d\eta \, Z[x_1 \eta] G_{n+1}(\eta\eta \, x_2 \cdots x_n)$$

$$+ ig \int d\eta \, Z[x_1 \eta] G_{n-1}(x_2 \cdots x_n), \qquad (8.2)$$

where, as usual, $[xy]$ is the free propagator. We have multiplied by a scaling factor Z, which disappears when taking matrix elements, because the wave functions also will carry suitable inverse powers of it.

Equations (8.2) are meaningless: it suffices to iterate them to find terms, such as $[xx]$ or $[xy]^2$, which are either just not defined at all or not distributions in S' [44]. Work with such a theory is particularly simple, because we find at most *pairs* of coincident variables in Eqs. (8.2). These equations we regularize with the $\hat{\delta}$ procedure (the σ procedure, or any other, would of course do as well). Define, for any m, the expressions

$$\overline{G}_{k+2m}^{(\eta_1 \cdots \eta_m)}(x_1 \cdots x_k \, \eta_1\eta_1 \cdots \eta_m\eta_m) =$$

$$= \theta(\lambda_1) \cdots \theta(\lambda_m) \int d\eta_1' \cdots \int d\eta_m' \, \hat{\delta}(\lambda_1, \eta_1 - \eta_1') \cdots$$

$$\hat{\delta}(\lambda_m, \eta_m - \eta_m')\overline{G}_{k+2m}(x_1 \cdots x_k \, \eta_1'\eta_1 \eta_2'\eta_2 \cdots \eta_m'\eta_m); \quad (8.3)$$

Eqs. (8.2) are thus replaced by their regularized counterparts

$$\overline{G}_n(x_1 \cdots x_n) = \sum_{J=2}^{n} \overline{Z}[\overline{x_1 x_J}]\overline{G}_{n-2}(x_2 \cdots x_{J-1} x_{J+1} \cdots x_n)$$

$$+ 3i\overline{g} \int d\eta \, \overline{Z}[\overline{x_1 \eta}]\overline{G}_{n+1}^{(\eta)}(\eta\eta \, x_2 \cdots x_n)$$

$$+ i\overline{g}_1 \int d\eta \, \overline{Z}[\overline{x_1 \eta}]\overline{G}_{n-1}(x_2 \cdots x_n) \qquad (8.4)$$

where we use for short the symbol $\int$ to denote

$$\int d\eta \, f(\eta x \cdots)g(\eta y \cdots) = \theta(\lambda) \int \int d\eta \, d\eta' f(\eta' x \cdots)g(\eta y \cdots)$$

$$\times \hat{\delta}(\lambda, \eta - \eta').$$

We want to show that the arbitrariness of the regularization procedure, exhibited through the arbitrary function $f(\lambda)$ incorporated into $\hat{\delta}(x, \lambda)$

(cf. (7.9)), is physically irrelevant; that is, changing $f(\lambda)$ into another such function does *not* change the values of the Green functions $G_n(x_1 \cdots x_n)$, which have all distinct points $x_1 \neq x_2 \neq \cdots \neq x_n$, provided that new "renormalized" parameters replace the old ones:

$$G(x_1 \cdots x_n \mid m, g, g_1) \equiv \bar{G}(x_1 \cdots x_n \mid \bar{m}, \bar{g}, \bar{g}_1) \tag{8.5}$$

and that for each choice of the function $f(\lambda)$ in (8.9) a change of parameters

$$m \to \bar{m}, \qquad g \to \bar{g}, \qquad g_1 \to \bar{g}_1, \qquad z \to \bar{z} \tag{8.6}$$

can be made that secures the validity of (8.5): the physics has *not* been changed by our regularization, which is therefore a *renormalization*. (8.5) and (8.6) just define, together, the renormalization. m is taken arbitrary, to ensure that any iterate of Eqs. (8.4) be equally well defined.

3. INFINITESIMAL RENORMALIZATIONS

The proof of this statement is simpler if done for an *infinitesimal* variation of $f(\lambda)$ in (7.9); this suffices, because there is a group property, as will be shown in the next chapters. This amounts to computing therefore the variation of all Green's functions in Eq. (8.4) when $f(\lambda) \to f(\lambda) + \delta f(\lambda)$ in (7.9); the only ones that manifestly vary are those with coincident pairs: (8.3) gives

$$\delta_f \bar{G}^{(\eta_1 \cdots \eta_m)}(x_1 \cdots x_k \, \eta_1 \eta_1 \cdots \eta_m \eta_m)$$

$$= \sum_{i=1}^{m} \theta(\lambda_1) \cdots \theta(\lambda_m) \hat{\delta}(\lambda_1, \eta_1 - \eta_1') \cdots \hat{\delta}(\lambda_{i-1}, \eta_{i-1} - \eta_{i-1}')$$

$$\times \{\delta_f \hat{\delta}(\lambda_i, \eta_i - \eta_i')\} \hat{\delta}(\lambda_{i+1}, \eta_{i+1} - \eta_{i+1}') \cdots$$

$$\times \hat{\delta}(\lambda_m, \eta_m - \eta_m') \bar{G}(x_1 \cdots x_k \, \eta_1 \eta_1' \cdots \eta_m \eta_m') \tag{8.7}$$

where

$$\delta_f \hat{\delta}(\lambda_i, \eta_i - \eta_i') = \frac{i}{\pi^2} \, \delta f(\lambda_i) \lambda_i (x^2 - i0^+)^{\lambda_i - 2} \tag{8.8}$$

The expressions (8.7) can be evaluated (this is, in fact, the typical advantage of our method) *by using the branching Eqs. (8.4) instead of the cumbersome standard graph-by-graph analysis.*

We easily find, after some simple manipulations, that if we set

$$a = 2\bar{Z}^2 \theta(\lambda_i) \int d\xi \int d\eta'_i \, \delta_f \hat{\delta}(\lambda_i, \eta_i - \eta'_i) [\overline{x\eta_i}] [\overline{x\eta_i}],$$

$$b = 2\bar{Z}^4 \theta(\lambda_i) \int d\xi_1 \int d\xi_2 \int d\eta'_i \, \delta_f \hat{\delta}(\lambda_i, \eta_i - \eta'_i) [\overline{\eta_i \xi_1}][\overline{\eta'_i \xi_2}] R(\xi_1 \xi_2),$$

$$c = \bar{Z}\theta(\lambda_i) \int d\eta'_i \, \delta_f \hat{\delta}(\lambda_i, \eta_i - \eta'_i) [\overline{\eta_i \eta'_i}] \tag{8.9}$$

with

$$R(\xi_1 \xi_2) = \theta(\lambda) \int d\xi \, \hat{\delta}(\lambda, \xi_1 - \xi) [\overline{\xi \xi_2}][\overline{\xi_2 \xi_1}], \tag{8.10}$$

the variation (8.7) is given by

$$\delta_f \bar{G}^{(\eta_1 \cdots \eta_m)}(x_1 \cdots x_k \, \eta_1 \eta_1 \cdots \eta_m \eta_m)$$

$$= 3iga \sum_{i=1}^{m} \bar{G}^{(\eta_1 \cdots \eta_{i-1} \eta_{i+1} \cdots \eta_m)}(x_1 \cdots x_k \, \eta_1 \eta_1 \cdots$$

$$\eta_{i-1} \eta_{i-1} \eta_i \eta_{i+1} \eta_{i+1} \cdots \eta_m \eta_m) + (c + 9(ig)^2 b)$$

$$\times \sum_{i=1}^{m} \bar{G}^{(\eta_1 \cdots \eta_{i-1} \eta_{i+1} \cdots \eta_m)}(x_1 \cdots x_k \, \eta_1 \eta_1 \cdots \eta_{i-1} \eta_{i-1}$$

$$\eta_{i+1} \eta_{i+1} \cdots \eta_m \eta_m)$$

$$+ \sum_{i=1}^{m} \sum_{k>i} \bar{G}^{(\eta_1 \cdots \eta_{i-1} \eta_{i+1} \cdots \eta_{k-1} \eta_{k+1} \cdots \eta_m)}(x_1 \cdots x_k$$

$$\eta_1 \eta_1 \cdots \eta_{i-1} \qquad \eta_{i+1} \eta_{i+1} \cdots \eta_{k-1} \eta_{k-1} \eta_{k+1} \eta_{k+1} \cdots \eta_m$$

$$\times F(\eta_i \eta_k) \tag{8.11}$$

with

$$F(\eta_i \eta_k) = Z^2 2\theta(\lambda_i) \int d\eta'_i [\overline{\eta_i \eta_k}] [\overline{\eta_i \eta_k}] \delta_f \hat{\delta}(\lambda_i, \eta_i - \eta'_i). \tag{8.12}$$

The meaning of a, b, and c is very simple:

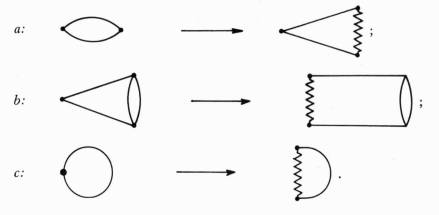

$a:$

$b:$

$c:$

Formulas such as (8.11) can be obtained, to any wanted m, from the branching equations (8.4).

We propose now to ascertain whether the changes brought into Eqs. (8.4) by the change of $\delta f(\lambda)$ in the regularization prescription can be compensated by suitable variations of the parameters $\overline{m}^2$, $\overline{g}$, $\overline{g}_1$, and $\overline{Z}$.

Taking into account that

$$\delta[\overline{xy}] = \frac{\partial[\overline{xy}]}{\partial(\overline{m}^2)} \delta(\overline{m}^2) = -\delta(\overline{m}^2)i \int d\eta \, [\overline{x\eta}] \, [\overline{\eta y}],$$

it can be shown (see [51] for the detailed proof) that the *total* variation of $\overline{G}$

$$\delta_{\text{tot}} \overline{G} = \delta_f \overline{G} + \frac{\partial \overline{G}}{\partial \overline{m}^2} \delta \overline{m}^2 + \frac{\partial \overline{G}}{\partial \overline{z}} \delta \overline{z} + \frac{\partial \overline{G}}{\partial \overline{g}} \delta \overline{g} + \frac{\partial \overline{G}}{\partial \overline{g}_1} \delta \overline{g}_1, \tag{8.13}$$

i.e. (8.5) holds, provided we take

$$\delta \overline{Z} = 0, \quad \text{therefore} \quad \overline{Z}' = \overline{Z}; \tag{8.14a}$$

$$\delta \overline{g} = 0, \quad \text{therefore} \quad \overline{g}' = \overline{g}; \tag{8.14b}$$

$$\delta \overline{m}^2 = 9\overline{z}\overline{g}^2 i, \quad \text{therefore} \quad (\overline{m}^2)' = \overline{m}^2 + 9ai\overline{Z}\overline{g}^2; \tag{8.14c}$$

$$\delta \overline{g}_1 = -3\overline{g}(c - 9b\overline{g}^2), \quad \text{therefore} \quad \overline{g}_1' = \overline{g}_1 - 3\overline{g}(c - 9b\overline{g}^2); \tag{8.14d}$$

for any equation, any iteration. This shows that the change δf in prescription, if accompanied by the change (8.14) in the unknown parameters $\overline{m}$, $\overline{g}$, $\overline{g}_1$, $\overline{Z}$ leaves the Green functions $G(x_1 \cdots x_n)$ $(x_1 \neq x_2 \cdots \neq x_n)$ and the equations unchanged.

Since $\overline{m}$, $\overline{g}$, $\overline{g}_1$ will be determined by experiment anyhow, it will make no difference to the physics to use any prescription whatever. This is all that is meant by renormalization.

4. PERTURBATIVE EXPANSIONS

It is also a straightforward matter, which will only be mentioned here, to prove that exactly *the same* is true for the perturbative expansions. The formal expressions

$$G_n(x_1 \cdots x_n) = \sum_{m,k=0}^{\infty} \frac{(ig)^k}{k!} \frac{(ig_1)^m}{m!} \int d\eta \int d\eta' \, [x_1 \cdots x_k$$

$$\times \eta_1 \eta_1 \eta_1 \cdots \eta_k \eta_k \eta_k \, \eta_1' \eta_2' \cdots \eta_m'], \tag{8.15}$$

when we take as elements of the hafnians $\bar{Z}[xy]$ and split each triple or pair of coincident points with the $\hat{\delta}$ regularization (for instance, the expression

$$\int d\eta \, f(\eta\eta\eta \, xy \cdots) \equiv \theta(\lambda)\theta(\lambda') \int\int\int d\eta \, d\eta' \, d\eta'' \, f(\eta\eta'\eta'' \, xy \cdots)$$
$$\times \hat{\delta}(\lambda, \eta - \eta')\hat{\delta}(\lambda', \eta - \eta')) \qquad (8.16)$$

can easily be seen, either through a graph-by-graph analysis or, more simply, by means of the branching equations, to yield *exactly* the same expansion we would have if we were starting from the renormalized branching equations (8.4). Likewise, a change of δf in the prescription will yield exactly *the same* change of renormalization (8.14) that was obtained with the branching equations.

5. A MODEL OF THE CHANGE IN ANALYTIC BEHAVIOR IN λ CAUSED BY RENORMALIZATION

Consider scalar electrodynamics; the propagators are

$$K_{NP} = K \begin{pmatrix} x_1 \cdots x_N \\ y_1 \cdots y_N \end{pmatrix} t_1 \cdots t_P \end{pmatrix} = \sum_{n(P)} \frac{\lambda^n}{n!} \int d\xi_1 \cdots d\xi_N$$

$$\times \begin{pmatrix} x_1 \cdots x_N \, \xi_1 \cdots \xi_n \\ y_1 \cdots y_N \, \xi_1 \cdots \xi_n \end{pmatrix} [\xi_1 \cdots \xi_n \, t_1 \cdots t_P], \qquad (8.17)$$

where $\lambda = ig$ and the sum extends over all positive values of n with the same parity of P. Take

$$[\xi\eta] = b = \text{constant}$$

to define our model [53, 54], so that the value of the hafnian is given simply by

$$[\xi_1 \cdots \xi_u \, t_1 \cdots t_p] = (p + n - 1)!!b^{(p+n)/2} \qquad (8.18)$$

where $p + n$ is even.

It can be shown that the terms

$$g_1\phi, \quad g_2\phi^2, \quad g_3\phi^3, \quad g_4\phi^4$$

must be added to the Lagrangian to have a renormalized theory. After *renormalization* we find (calculations are omitted)

$$K_{NP} = (2\pi b)^{-1/2} \int_{-\infty}^{+\infty} \exp\left[-\frac{\tau^2}{2b}\right] \tau^P \exp\left[\frac{i\Omega}{16\pi^2}(m - g\tau)^4\right]$$

$$\times \log \frac{(m - g\tau)^2}{m_0^2}\right] \exp[i\Omega(g_1\tau + g_2\tau^2 + g_3\tau^3 + g_4\tau^4)]$$

$$\times \begin{pmatrix} x_1 \cdots x_N \\ y_1 \cdots y_N \end{pmatrix}_{m - g\tau} d\tau. \tag{8.19}$$

We observe that if we set $g = g_1 = g_2 = g_3 = g_4 = 0$, the propagator K_{NP} reduces directly to the free propagator. It is then easily ascertained that K_{NP} expressed in the form (8.19) satisfies all the branching equations.

This example is interesting because it shows that after renormalization all propagators and Green's functions *are of Hadamard's class 2* [36] and have therefore an essential singularity at $\lambda = 0$, whereas the perturbative series (8.17), regularized by assuming a finite integration volume and defining the free-fermion propagator such that $|(\xi\eta)| \leqslant M$, is *analytic in g at the origin* with a radius of convergence that is at least finite. In fact, assuming for simplicity that $P = 0$ and making use of Hadamard's majoration for the determinant, we have that the nth term of the series (8.17) is majorated in modulus by

$$\frac{|g|^n}{n!}(n - 1)!!b^{(n/2)}\Omega^n 4^{(N+n-1)}M^{N+n}(N+n)^{(N+n)/2}$$

$$= \frac{|g|^{2m}}{(2m)!}(2m - 1)!!4^{(N+2m-1)}b^m \Omega^{2m} M^{N+2m}$$

$$\times (N + 2m)^{(M/2)+m}, \qquad n = 2m, \tag{8.20}$$

which is the general term of a power series in g having a finite radius of convergence independent of N. (This radius could also be made infinite by taking a finite number of fermion modes. See Section 4 of Chapter 5.)

The analytic behavior of the exact solution with respect to g at the origin is therefore *profoundly* different from that of the regularized perturbative expansions.

Chapter 9

Renormalization

1. FROM REGULARIZATION TO RENORMALIZATION

The two preceding chapters have presented some specific regularizations (two of them were defined in order to emphasize the lack of uniqueness a priori inherent in the choice of any such procedure) and have shown how the branching equations and their formal perturbative solutions thereby become mathematically well defined and meaningful in a typical field theory (chosen superrenormalizable for maximal simplicity). The reader not specifically interested in the subject of renormalization may stop at this point, if he will only believe our statement (proved years ago in [15, 16] that this is also true for all renormalizable theories (according to the classic definition of Dyson [55]*): if he just replaces the formal expressions of Part II with those immediately derivable from them by adopting the σ or the $\hat{\delta}$ regularization, he can, in any computation, completely forget about the problem of renormalization, because all computations are certain to yield correctly renormalized quantities.†

The aim of the present chapter is to indicate how and why this happens. Proofs will be given only when the work referred to has not yet been published (these proofs are all due to M. Marinaro); we shall otherwise refer the reader to earlier references, trying to keep the presentation as uncumbersome as possible.

The crux of the matter is that ill-defined products of distributions (in the older parlance, divergent integrals) soon appear when branching equations are iterated or formal expansions considered. The alternative

---

*A theory described by a Lagrangian $\mathscr{L}(x)$ is renormalizable if $\mathscr{L}(x)$ is a sum of products of fields, each of dimension $d \leqslant 4$.

† All extant methods solve the problem of renormalization theoretically, but in ways which are, at best, awkward for computation. A quite different approach, which may offer better perspectives in this respect, is indicated by the author in "Generalized Integration Procedure for Divergent Integrals", Nuovo Cimento, in press.

is obvious: either our equations or Lagrangians are not good enough
(e.g., because they do not contain mention of a fundamental length or
of the gravitational field), or, if we choose to go ahead without changing
the theory, we must *add to it* some prescription to define somehow
those products of distributions (or, equivalently, to extract convergent
values out of formally divergent integrals). Of course, not all regularizations or field theories can be "good." Of the first we must ask that they
preserve the properties of unitarity, relativistic invariance, causality, and
permanence (by which is meant, e.g., that if some integral is already
convergent, its value must not be altered by the regularization: this
mathematical principle clearly must dominate, as the only means to
avoid chaos, every operation of continuation or extension). Of the
second, we ask that their Lagrangians be "complete" to start with.*

It so happens that among the infinitely many possible Lagrangians
that can couple spinor and boson fields, a very few polynomials of very
low degree satisfy this requirement: all corresponding theories (among
them, electrodynamics!) are called renormalizable; they share the remarkable property that only for them can we define *regularizations that do
not alter their physical content* (i.e., that are "good" in our sense). Since
we study here only theories of this class, we adopt the standard practice
of calling renormalization a "good" regularization (see, however, Section 4
for further comments on this point).

The statement that "the physics is unchanged" can be conveniently
rephrased as follows. Recall (e.g., from Lie group theory) the definition
of *essential parameter a* in a function $f(x; a)$: thus, if a is essential in
$f(x; a)$, a and b cannot *both* be essential in $f(x; a + b)$. Then, if the
Lagrangian of a renormalizable theory contains as *essential parameters*
some masses $m_1, \ldots, m_k$ and some coupling constants $g_1, \ldots, g_l$, whatever parameters may be introduced by a renormalization *will always be
inessential*, that is, will combine with the (a priori unknown) $m_1, \ldots, m_k$,
$g_1, \ldots, g_l$ into some new expressions $\bar{m}_1, \ldots, \bar{m}_k, \bar{g}_1, \ldots, \bar{g}_l$ (whose
values are also unknown and must be ascertained from experiment).

We have chosen this mode of presentation because there has been an
ever increasing variety of methods for defining renormalization, which

---

*The classic example is that of the nucleon–meson coupling $g\bar{\psi}\gamma^5\psi\phi$ [56, 57] which becomes
complete only after addition of the Matthews term $g_1\phi^4$; this question is discussed in detail in
[58], where it is shown that our combinatoric formalism provides answers as well to all such
questions; for simplicity's sake, we assume it already solved here, and confine our work to the
standard case of the neutral meson field with self-coupling $g\phi^4$.

are all proved or claimed "not to change the physics," without any obvious connection among them. Students find this situation bewildering and demanding, because quite often the inadequacy of the combinatoric tools conspires with the formidability of the analytic apparatus* to give proofs that easily exceed a hundred pages. We propose to show here, with a careful discrimination between combinatoric and distribution theoretical properties, that if we work with propagators and in x space, all that pertains to combinatorics can be handled with the same ease with which in Chapter 3 the formal perturbative expansions were obtained. These methods also permit us to recognize the full equivalence between distinct renormalization procedures, however different their *prima facie* looks may be.

2. METHODS OF PROOF

The following problems therefore arise.

(a) Given a regularization, we have to make sure that it does not alter the physics (i.e. that it is a renormalization).

(b) We have to exhibit the connections between any two distinct renormalizations.

Let us start with problem (a). We have studied it with two different methods, each of which has distinct formal advantages and drawbacks. We call them *finite* and *infinitesimal*, respectively.

The *finite method* is based on the Dyson criterion [55] that all numerical values of S- or U-matrix elements between any initial and final states must stay unchanged under renormalization. Denoting symbolically with $\int$ and $\int$, respectively, computation made with the formal theory and with the theory regularized in some consistent way (e.g., as indicated in Chapter 7), this criterion amounts to defining "renormalization" as a regularization for which in terms of propagators (recall that we restrict our discussion to the $g\phi^4$ theory)

$$K_{2N_0}(x_1 \cdots x_{2N_0}; m, g, \textstyle\int) = Z^{N_0}(m, g, \textstyle\int) A(m, g, \textstyle\int)$$
$$\bar{K}_{2N_0}(x_1 \cdots x_{2N_0}; \bar{m}(m, g, \textstyle\int), \bar{g}(m, g, \textstyle\int)$$
$$(9.1)$$

---

* Such apparatus is employed also with a view to proving the *existence* of solutions: but this goal is quite another problem, and mixing it with a discussion of renormalization would only confuse the present issue.

In (9.1), Z denotes a change of scale*; the relation between the formal mass and charge m and g and $\overline{m}$ and $\overline{g}$ is discussed [15, 16]. We forego here a discussion of the finite method, as we have already treated it exhaustively in a number of works [15, 16, 58]. We shall only indicate some of its main features, and at the end of this chapter supply a "notational key" to facilitate the task for the interested reader. In short, relation (9.1) is proved with this method through, and for all, branching equations of *either type;* it acts by eliminating all divergences that can arise *at a single x-space point* (and by iteration at all points; cf. the Bogolubov program); the renormalized perturbative expansions are verified a posteriori to satisfy the renormalized equations, but the direct verification of (9.1) in expanded form is rather involved.

The *infinitesimal method* is also inspired by relation (9.1); this in fact indicates that whatever parameters are introduced into K_{2N_0} when computing with a renormalization, they are *inessential* because absorbed into $\overline{m}$ and $\overline{g}$ (besides Z and A). Denote with R the set of all such parameters; we now ask that

$$\overline{K}_{2N_0}(x; \overline{m}, \overline{g}, R) = Z^{N_0}(\overline{m}, \overline{g}, R)A(\overline{m}, g, R)\overline{\overline{K}}_{2N_0}(x; \overline{\overline{m}}(\overline{m}, \overline{g}, R),$$

$$\overline{\overline{g}}(\overline{m}, \overline{g}, R)) \tag{9.2}$$

where for formal simplicity we still use the letters $\overline{m}$, $\overline{g}$, but it must be clear that we are thus indicating some parameters that do not have the same meaning as in (9.1).

A renormalization is now defined as a regularization for which (9.2) holds. The infinitesimal method will be applied in the next sections to the $g\phi^4$ theory, and will be seen to have the advantage of enabling us to handle with equal ease branching equations of both types and perturbative expansions, and the disadvantage of not allowing an immediate comparison with the formal theory. (As regards problem (b), it is solved equally well by both methods, as will be shown in the sequel.)

In conclusion, it must be emphasized that Eqs. (9.1) and (9.2) are only two ways of saying the same thing. Requiring that they be satisfied ensures the invariance of all matrix elements, as well as of the *form* of all equations and expansions (for this reason we call such procedures "form-invariant renormalization"). Unitarity, causality, and Lorentz

---

* This notation *differs* from that used in Chapter 8 with the $g\phi^3$ theory: we do this in order to facilitate comparison with our past works on the subject; the difference is otherwise irrelevant.

invariance are thereby satisfied as well; it is essential to point out, however, that "the physics is unchanged" only if the physical masses and coupl
constants stay real (if necessary, positive). Unlike what happens with
other formalisms, in which the physical masses and charges are introduced
through counterterms so as to satisfy the latter requirement, in our case
this condition must be *added* after computation. It will be shown in the
last section and in the next chapter that new interesting possibilities
(mass spectra, i.e., inequivalent representations) do arise, at least in
simple models.

3. THE $g\phi^4$ THEORY

The $g\phi^4$ theory is described formally by the expansion (5.27) and
Eqs. (5-28)-(5.30), which we repeat here, for convenience, in terms of
the propagators

$$K_{2N_0} \equiv K(x_1 \cdots x_{2N_0}) = \sum_{N=0}^{\infty} \frac{(-ig)^N}{N!} \int d\xi_1 \cdots \int d\xi_N$$

$$[x_1 \cdots x_{2N_0} \, \xi_1 \xi_1 \xi_1 \xi_1 \xi_2 \xi_2 \xi_2 \xi_2 \cdots \xi_N \xi_N \xi_N \xi_N], \tag{9.3}$$

$$\tilde{K}(x_1 \cdots x_{2N_0}) = \sum_{h=2}^{2N_0} [x_1 x_h] \tilde{K}_{2N_0 - 2}(x_2 \cdots x_{h-1} x_{h+1} \cdots x_{2N_0})$$

$$-4ig \int d\xi_1 [x_1 \xi_1] \tilde{K}_{2N_0+2}(x_2 \cdots x_{2N_0} \xi_1 \xi_1 \xi_1), \tag{9.4}$$

$$\partial_\mu \tilde{K}(x_1 \cdots x_{2N_0}) = \frac{-i}{2} \int \tilde{K}_{2N_0+2}(x_1 \cdots x_{2N_0} \xi_1 \xi_1) \, d\xi_1,$$

$$\partial_g \tilde{K}(x_1 \cdots x_{2N_0}) = -i \int d\xi_1 \tilde{K}_{2N_0+4}(x_1 \cdots x_{2N_0} \xi_1 \xi_1 \xi_1 \xi_1). \tag{9.5}$$

We begin by regularizing these equations with the $\hat{\delta}$ procedure. This
yields

$$\bar{K}_{2N_0} \equiv \bar{K}_{2N_0}(x, \bar{m}, \bar{g}, R)$$

$$= \sum_N \frac{(-i\bar{g})^N}{N!} \theta(\lambda) \int dR_N \, \Gamma_1^{(4)} \cdots \Gamma_N^{(4)}$$

$$\times [x_1 \cdots x_{2N_0} \, \xi_1^{(1)} \cdots \xi_1^{(4)} \cdots \xi_N^{(1)} \cdots \xi_N^{(4)}] \tag{9.6}$$

where $[xy] = \Delta_F(x - y, \bar{m})$, $\bar{m}$ and $\bar{g}$ represent a mass and a charge,

$\Gamma_i^{(4)}$, $\int dR_N$ are obtained from (7.28) by putting $i_1 = i_2 = \cdots = i_k = 4$, and $\theta(\lambda) = \theta(\lambda_1^{(1)}) \cdots \theta(\lambda_1^{(4)}) \cdots \theta(\lambda_n^{(1)}) \cdots \theta(\lambda_n^{(4)})$, the symmetrization of the limiting procedures, as a consequence of the symmetry of the integrand, is superfluous.

With the notation (7.27), (9.6) reduces to

$$\bar{K}_{2N_0} = \sum_N \frac{(-i\bar{g})^N}{N!} \int d\xi_1 \cdots \int_{\xi_1 \cdots \xi_{N-1}} d\xi_N [x_1 \cdots x_{2N_0}$$

$$\xi_1\xi_1\xi_1\xi_1 \cdots \xi_N\xi_N\xi_N\xi_N]. \tag{9.7}$$

The individual terms of this series are now well defined (as is shown in Chapter 7; nothing can or must be said here regarding its convergence). In the same way, Eqs. (9.4), (9.5) become, after regularization,

$$\bar{K}(x_1 \cdots x_{2N_0}) = \sum_i [\overline{x_1 x_i}] \bar{K}_{2N_0 - 2}(x_2 \cdots x_{i-1} x_{i+1} \cdots x_{2N_0})$$

$$-4i\bar{g} \int d\xi\, [\overline{x_1 \xi}] \bar{K}^{(\xi)}_{2N_0+2}(\xi\xi\xi x_2 \cdots x_{2N_0}), \tag{9.8}$$

$$\frac{\partial}{\partial \bar{g}} \bar{K}(x_1 \cdots x_{2N_0}) = -i \int d\xi_1\, \bar{K}^{(\xi_1)}_{2N_0+4}(x_1 \cdots x_{2N_0} \xi_1\xi_1\xi_1\xi_1),$$

$$\frac{\partial}{\partial \bar{m}^2} \bar{K}(x_1 \cdots x_{2N_0}) = -\frac{i}{2} \int d\xi_1 \bar{K}^{(\xi_1)}_{2N_0+2}(x_1 \cdots x_{2N_0} \xi_1\xi_1) \tag{9.9}$$

where we write, for short, $\bar{K}$ in place of $\bar{\bar{K}}$, $(\Box - m^2)[\overline{xy}] = i\delta(x - y)$,

$$(\Box - \bar{m}^2)[\overline{xy}] = i\delta(x - y),$$

$$\int d\eta\, f(\eta x \cdots) g(\eta y \cdots) \overset{\text{def.}}{\equiv} \theta(\lambda) \int d\eta \int d\eta' f(\eta x \cdots) g(\eta' y \cdots)$$

$$\times \hat{\delta}(\eta - \eta', \lambda)$$

$$\bar{K}^{(\xi_1 \cdots \xi_K)}(\underbrace{\xi_1 \cdots \xi_1}_{i_1} \cdots \underbrace{\xi_K \cdots \xi_K}_{i_K})$$

$$= \theta(\lambda) \int d\xi_1^{(1)} \cdots \int d\xi_1^{(i_1)} \cdots \int d\xi_K^{(1)} \cdots \int d\xi_K^{(i_K)}$$

$$\times \Gamma_1^{(i_1)} \cdots \Gamma_K^{(i_K)} \bar{K}(\xi_1^{(1)}\xi_1^{(2)} \cdots \xi_1^{(i_1)} \cdots \xi_K^{(1)} \cdots \xi_K^{(i_K)}). \tag{9.10}$$

In order to prove that the $\hat{\delta}$ regularization is a renormalization, that is, that Eq. (9.2) is satisfied, we have to compute the variation of $\bar{K}_{2N_0}$ under a change of f in Eqs. (7.9) and show that

$$\delta_R \bar{K}_{2N_0} \equiv \delta_f \bar{K}_{2N_0} = [N_0 Z^{-1} \delta_f Z + A^{-1} \delta_f A] Z^{N_0} A \bar{K}_{2N_0}$$

$$+ Z^{N_0} A \left[\frac{\partial \bar{\bar{K}}_{2N_0}}{\partial \bar{m}} \delta_f \bar{\bar{m}} + \frac{\partial \bar{\bar{K}}_{2N_0}}{\partial \bar{\bar{g}}} \delta_f \bar{\bar{g}} \right]$$

or expressed through $\bar{K}$

$$\delta_f \bar{K}_{2N_0} = \bar{K}_{2N_0} \left\{ A^{-1} \left[\delta_f A - \frac{1}{Q} \frac{\partial A}{\partial \bar{m}} \left(\frac{\partial \bar{\bar{g}}}{\partial \bar{g}} \delta_f \bar{\bar{m}} - \frac{\partial \bar{\bar{m}}}{\partial \bar{g}} \delta_f \bar{\bar{g}} \right) \right. \right.$$

$$\left. + \frac{1}{Q} \frac{\partial A}{\partial \bar{g}} \left(\frac{\partial \bar{\bar{g}}}{\partial \bar{m}} \delta_f \bar{\bar{m}} - \frac{\partial \bar{\bar{m}}}{\partial \bar{m}} \delta_f \bar{\bar{g}} \right) \right] \right\} + N_0 \bar{K}_{2N_0} \left\{ Z^{-1} \left[\delta_f Z \right. \right.$$

$$\left. \left. + \frac{1}{Q} \frac{\partial Z}{\partial \bar{g}} \left(\frac{\partial \bar{\bar{g}}}{\partial \bar{m}} \delta_f \bar{\bar{m}} - \frac{\partial \bar{\bar{m}}}{\partial \bar{m}} \delta_f \bar{\bar{g}} \right) - \frac{1}{Q} \frac{\partial Z}{\partial \bar{m}} \left(\frac{\partial \bar{\bar{g}}}{\partial \bar{g}} \delta_f \bar{\bar{m}} - \frac{\partial \bar{\bar{m}}}{\partial \bar{g}} \delta_f \bar{\bar{g}} \right) \right] \right\}$$

$$+ \frac{1}{Q} \left(\frac{\partial \bar{\bar{g}}}{\partial \bar{g}} \delta_f \bar{\bar{m}} - \frac{\partial \bar{\bar{m}}}{\partial \bar{g}} \delta_f \bar{\bar{g}} \right) \frac{\partial \bar{K}_{2N_0}}{\partial \bar{m}} - \frac{1}{Q} \left(\frac{\partial \bar{\bar{g}}}{\partial \bar{m}} \delta_f \bar{\bar{m}} - \frac{\partial \bar{\bar{m}}}{\partial \bar{m}} \delta_f \bar{\bar{g}} \right)$$

$$\times \frac{\partial \bar{K}_{2N_0}}{\partial \bar{g}} \tag{9.11}$$

where

$$Q = \frac{\partial \bar{\bar{m}}}{\partial \bar{m}} \frac{\partial \bar{\bar{g}}}{\partial \bar{g}} - \frac{\partial \bar{\bar{m}}}{\partial \bar{g}} \frac{\partial \bar{\bar{g}}}{\partial \bar{m}}.$$

We start with the perturbative expansion and calculate

$$\delta_f \bar{K}_{2N_0} = \sum_{N=1}^{\infty} \frac{(-i\bar{g})^N}{N!} \sum_{i=1}^{N} \theta(\lambda) \int dR_N \, \Gamma_1^{(4)} \cdots \Gamma_{i-1}^4$$

$$\times (\delta_f \Gamma_i^{(4)}) \Gamma_{i+1}^{(4)} \cdots \Gamma_N^{(4)} [x_1 \cdots x_{2N_0} \xi_1^{(1)} \cdots \xi_1^{(4)} \cdots] \tag{9.12}$$

where

$$\delta_f \Gamma_i = \delta_f \left[\prod_{l=1}^{4} \hat{\delta}(\xi_i^l - \xi_i, \lambda_i^l) \right].$$

The integrand in (9.12) is symmetric with respect to all the variables except the ith; it is therefore convenient to order the limiting procedures in such a way as to perform first the ith limit. To do this we use the relation

$$\sum_{i=1}^{N} \theta(\lambda_1) \cdots \theta(\lambda_i) \cdots \theta(\lambda_N) F_i(1, 2, \ldots, N)$$

$$= \sum_{i=1}^{N} \sum_{k=0}^{N-i} \binom{N-i}{k} \theta(\lambda_1) \cdots \theta(\lambda_{i-1}) \theta(\lambda_{i+k+1}) \cdots \theta(\lambda_N)$$

$$\times \theta_{\lambda_i, \lambda_{i+1}, \ldots, \lambda_{i+k}} F_i(1, 2, \ldots, N) * \tag{9.13}$$

where, denoting $[A, B]$ the commutator, we have called

$$\theta_{\lambda_i, \lambda_{i+1}, \ldots, \lambda_k} = [\theta_{\lambda_i, \lambda_{i+1}, \ldots, \lambda_{k-1}}, \theta(\lambda_k)], \tag{9.14a}$$

$$\theta_{\lambda_i} \equiv \theta(\lambda_i) = \theta(\lambda_i^{(1)}) \cdots \theta(\lambda_i^{(4)}). \tag{9.14b}$$

$F_i(1, 2, \ldots)$ is a function symmetric with respect to all variables except i.

The expression (9.13) can be put in the form

$$\sum_{i=1}^{N} \theta(\lambda_1) \cdots \theta(\lambda_i) \cdots \theta(\lambda_N) F_i(1, 2, \ldots, N)$$

$$= \sum_{s=1}^{N} \binom{N}{s-1} \theta(\lambda_2) \cdots \theta(\lambda_s) \theta_{\lambda_1, \lambda_{s+1}, \ldots, \lambda_N}$$

$$\times F_1(1, 2, \ldots, N). \tag{9.15}$$

In conclusion

$$\delta_f \bar{K}_{2N_0} = \sum_{N=1}^{\infty} \frac{(-i\bar{g})^N}{N!} \sum_{s=1}^{N} \binom{N}{s-1} \theta(\lambda_2) \cdots \theta(\lambda_s) \theta_{\lambda_1, \lambda_{s+1}, \ldots, \lambda_N}$$

$$\int dR_N (\delta_f \Gamma_1) \Gamma_2 \cdots \Gamma_N [x_1 \cdots x_{2N_0} \xi_1^{(1)} \cdots \xi_1^{(4)} \cdots$$

$$\times \xi_N^{(1)} \cdots \xi_N^{(4)}]. \tag{9.16}$$

---

* Clearly the left-hand side of (9.13) is $\theta(\lambda_1) \cdots \theta(\lambda_{i-1}) \theta(\lambda_{i+1}) \cdots \theta(\lambda_k) \theta(\lambda_i)$ for $K = 0$ and $\theta(\lambda_1) \cdots \theta(\lambda_{i-1}) \theta_{\lambda_i, \lambda_{i+1}, \ldots, \lambda_N}$ for $K = N - 1$.

Reversing the order of the summations we have

$$\delta_f \bar{K}_{2N_0} = \sum_{s=1}^{\infty} \sum_{N=s}^{\infty} \frac{(-i\bar{g})^N}{N!} \binom{N}{s-1} \theta(\lambda_2) \cdots \theta(\lambda_s) \theta_{\lambda_1, \lambda_{s+1}, \ldots, \lambda_N}$$

$$\times \int dR_N (\delta_f \Gamma_1) \Gamma_2 \cdots \Gamma_N [x_1 \cdots x_{2N_0}$$

$$\xi_1^{(1)} \cdots \xi_1^{(4)} \cdots \xi_N^{(1)} \cdots \xi_N^{(4)}]. \tag{9.17}$$

Keeping in mind Eqs. (3.12), (3.12′), we can write

$$\delta_f \bar{K}_{2N_0} = \sum_{s=1}^{\infty} \frac{(-i\bar{g})^{s-1}}{(s-1)!} \theta(\lambda_2) \cdots \theta(\lambda_s) \int dR_{s-1} \Gamma_2 \cdots \Gamma_s$$

$$\times \sum_{s=1}^{N_0+2s-2} \sum_{C\rho} [u_1 \cdots u_{2\rho}] R(\overset{\circ}{w}_1 \cdots \overset{\circ}{w}_\sigma) \tag{9.18}$$

$\Sigma_{C\rho}$ denotes summation over all possible ways of extracting in natural order 2ρ terms $u_1 \cdots u_{2\rho}$ from the sequence

$$\{x_1 \cdots x_{2N_0} \xi_2^{(1)} \cdots \xi_2^{(4)} \cdots \xi_s^{(1)} \cdots \xi_s^{(4)}\}$$

$$R(\overset{\circ}{w}_1 \cdots \overset{\circ}{w}_\sigma) = \sum_{h=0}^{\infty} \frac{(-i\bar{g})^{h+1}}{(h+1)!} \theta_{\lambda_1, \lambda_2, \ldots, \lambda_{h+1}} \int dR_{h+1}$$

$$\times (\delta_f \Gamma_1) \Gamma_2 \cdots \Gamma_{h+1} [\overset{\circ}{w}_1 \cdots \overset{\circ}{w}_\sigma$$

$$\xi_1^{(1)} \cdots \xi_1^{(4)} \cdots \xi_{h+1}^{(1)} \cdots \xi_{h+1}^{(4)}]. \tag{9.19}$$

We want to analyze the quantities $R(\overset{\circ}{w}_1, \ldots, \overset{\circ}{w}_\sigma)$. To this end we use the fact that, given n operators $A_1, \ldots, A_n$, the expression

$$[\ldots, [A_1, A_2], A_3], \ldots, A_n] = 0 \tag{9.20}$$

if the operators can be divided in two sets $(A_{i_1}, \ldots, A_{i_s})$, $(A_{i_{s+1}}, \ldots, A_{i_{s+k}})$ such that the operators of a set commute with the operators of the other. Keeping in mind the definition (9.14) of $\theta_{\lambda_1, \ldots, \lambda_n}$, it follows from (9.20) that

$$\theta_{\lambda_1, \lambda_2 \ldots \lambda_k} f_1(1, i_1, \ldots, i_l) f_2(i_{l+1}, \ldots, i_k) = 0; \tag{9.21}$$

that is, contributions to R come from connected graphs only. Furthermore, a deeper analysis shows that $\theta_{\lambda_1, \lambda_2, \ldots, \lambda_l}$ yields a nonvanishing result only when acting on expressions that correspond to graphs with at most four external legs (the only ones that exhibit "superficial divergence," i.e., overall divergence according to power counting).

Proof. This property of $\theta_{\lambda_1,\lambda_2,\ldots,\lambda_l}$ is proved most simply by the induction method. Denote for short always with $f(1, 2, \ldots, n)$ any expression corresponding to graphs with n points $(1, 2, \ldots, n)$ and an arbitrary number of external legs, and observe that

$$\theta_{\lambda_1,\lambda_2} f(1, 2) = 0$$

for $f(1, 2)$ corresponding to graphs with more than four legs. Then, suppose that $\theta_{\lambda_1,\ldots,\lambda_k} f(1, \ldots, k) = 0$ if $f(1, \ldots, k)$ corresponds to graphs with k points and more than four legs. We must show that

$$\theta_{\lambda_1,\ldots,\lambda_{k+1}} f(1, \ldots, k + 1) = 0$$

if $f(1, \ldots, k + 1)$ corresponds to graphs with $k + 1$ points and more than four legs. Divide all graphs $(1, \ldots, k)$ into three classes

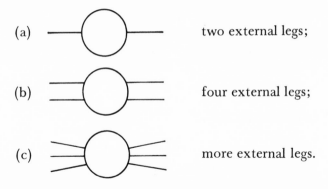

(a)   two external legs;

(b)   four external legs;

(c)   more external legs.

The graphs $(1, \ldots, k + 1)$ can be obtained from $(1, \ldots, k)$ by connecting the legs of point $k + 1$ ($4{:}\times$ or $2{:}\bigcirc$) in all possible ways to the external legs of $(1, 2, \ldots, k)$. From (a) it is not possible to obtain connected graphs $(1, 2, \ldots, k + 1)$ with more than four legs; from (b) the only graphs with more than four legs are of the type

and in this case $\theta(\lambda_{k+1})\theta_{\lambda_1,\ldots,\lambda_k} = \theta_{\lambda_1,\ldots,\lambda_k}\theta(\lambda_{k+1})$; therefore the application of $\theta_{\lambda_1,\ldots,\lambda_{k+1}}$ to these graphs gives zero. We are left with class (c), for which, according to our hypothesis, we have

$$\theta_{\lambda_1,\ldots,\lambda_k} f(1,\ldots,k) = 0$$

and therefore

$$\theta_{\lambda_1,\ldots,\lambda_k\lambda_{k+1}} f \equiv \theta_{\lambda_1,\ldots,\lambda_k} \theta_{\lambda_{k+1}} f$$

can be different from zero only if $f(1,\ldots,k+1)$ is superficially divergent, that is, if the graph $(1, 2, \ldots, k+1)$ has at most four external legs.

In conclusion, it becomes obvious that

$$\theta_{\lambda_1,\ldots,\lambda_n} f(1,\ldots,n) = 0 \tag{9.22}$$

if $f(1,\ldots,n)$ corresponds to a graph with more than four external legs. ∎

From (9.21) and (9.22) it follows that $R(\overset{\circ}{w}_1,\ldots,\overset{\circ}{w}_\sigma)$ is equal zero for $\sigma > 4$ and

$$R(0) = \sum_{h=0}^{\infty} \frac{(-i\bar{g})^{h+1}}{(h+1)!} \theta_{\lambda_1,\lambda_2,\ldots,\lambda_{h+1}} \int dR_{h+1}\, (\delta_f \Gamma_1) \cdots \Gamma_{h+1}$$

$$[\xi_1^{(1)} \cdots \xi_1^{(4)} \cdots \xi_{h+1}^{(1)} \cdots \xi_{h+1}^{(4)}] = R_0 \tag{9.23}$$

$$R(\overset{\circ}{w}_1\overset{\circ}{w}_2) = R_2^{(1)} \int d\xi [\overset{\circ}{w}_1 \overset{\circ}{w}_2\, \xi\xi] + \tfrac{1}{2} R_2^{(2)}$$

$$\times \int d\xi ([w_1\,\xi]\,\delta(\xi - w_2) + [w_2\,\xi]\,\delta(\xi - w_1)), \tag{9.24}$$

$$R(\overset{\circ}{w}_1\overset{\circ}{w}_2\overset{\circ}{w}_3\overset{\circ}{w}_4) = R_4 \int d\xi [\overset{\circ}{w}_1\overset{\circ}{w}_2\overset{\circ}{w}_3\overset{\circ}{w}_4\,\xi\xi\xi\xi] \tag{9.25}$$

where

$$R_2^{(1)} = \sum_{h=0}^{\infty} \frac{(-i\bar{g})^{h+1}}{(h+1)!} \theta_{\lambda_1,\lambda_2,\ldots,\lambda_{h+1}} \int d\xi_1^{(1)} \cdots \int d\xi_1^{(4)}$$

$$\times \int dR_h(\delta\Gamma_1)\Gamma_2 \cdots \Gamma_{h+1} \{ \underset{(2)}{\Sigma} \, [\xi_1^{(1)} \cdots \xi_1^{(4)} \cdots \xi_{h+1}^{(1)} \cdots$$

$$\xi_{h+1}^{(4)}] - \bar{m}^2 h(h-1)(\xi_2 - \xi_3)^2 [\xi_1^{(1)} \cdots \xi_1^{(4)}\xi_2^{(2)} \cdots$$

$$\xi_2^{(4)}\xi_3^{(2)} \cdots \xi_3^{(4)} \cdots \xi_4^{(1)} \cdots \xi_4^{(4)} \cdots \xi_{h+1}^{(1)} \cdots \xi_{h+1}^{(4)}]$$

$$- 2\bar{m}^2 h(\xi_2 - \xi_1)^2 [\xi_1^{(2)} \cdots \xi_1^{(4)}\xi_2^{(2)} \cdots \xi_2^{(4)}\xi_3^{(1)} \cdots$$

$$\xi_3^{(4)} \cdots \xi_{h+1}^{(4)} \cdots \xi_{h+1}^{(4)}]\}$$

$$R_2^{(2)} = \sum_{h=0}^{\infty} \frac{(-i\bar{g})^{h+1}}{(h+1)!} \theta_{\lambda_1, \lambda_2, \ldots, \lambda_{h+1}} \int d\xi_1^{(1)} \cdots \int d\xi_1^{(4)}$$

$$\times \int dR_h(\delta\Gamma_1)\Gamma_2 \cdots \Gamma_{h+1}\{-2h(h-1)(\xi_3 - \xi_2)^2$$

$$\times [\xi_1^{(1)} \cdots \xi_1^{(4)}\xi_2^{(2)} \cdots \xi_2^{(4)}\xi_3^{(2)} \cdots \xi_3^{(4)} \cdots \xi_{h+1}^{(1)} \cdots$$

$$\xi_{h+1}^{(4)}] - 4h(\xi_2 - \xi_1)^2[\xi_1^{(2)} \cdots \xi_1^{(4)}\xi_2^{(2)} \cdots \xi_2^{(4)} \cdots$$

$$\xi_{h+1}^{(1)} \cdots \xi_{h+1}^{(4)}]\}.$$

$$R_4 = \sum_{h=0}^{\infty} \frac{(-i\bar{g})^{h+1}}{(h+1)!} \theta_{\lambda_1, \lambda_2, \ldots, \lambda_{h+1}} \int d\xi_1^{(1)} \cdots \int d\xi_1^{(4)}$$

$$\times \int dR_h(\delta\Gamma_1)\Gamma_2 \cdots \Gamma_{h+1} \sum_{(4)} [\xi_1^{(1)} \cdots \xi_1^{(4)} \cdots \xi_{h+1}^{(1)} \cdots \xi_{h+1}^{(4)}],$$

and $\Sigma_{(q)}$ denotes summation over all hafnians obtainable from $[\xi_1^{(1)} \cdots \xi_1^{(4)} \cdots \xi_{h+1}^{(1)} \cdots \xi_{h+1}^{(4)}]$ by deleting any q variables from them. The computation of (9.23)–(9.25) follows the same procedure adopted in Appendix B of [52]). Substitution of (9.23)–(9.25) in (9.18) gives

$$\delta_f \bar{K}_{2N_0} = R_0 \bar{K}_{2N_0} + N_0 R_2^{(2)} \bar{K}_{2N_0} + R_2^{(1)} \sum_{s=0}^{\infty} \frac{(-ig)^s}{s!} \theta_{\lambda_1} \cdots \theta_{\lambda_s}$$

$$\times \int d\xi \int dR_s \, \Gamma_1 \cdots \Gamma_s [x_1 \cdots x_{2N_0} \xi_1^{(1)} \cdots \xi_1^{(4)} \cdots$$

$$\times \xi_s^{(1)} \cdots \xi_s^{(4)} \xi\xi] + (R_4 - 2ig R_2^{(2)})$$

$$\times \sum_{s=0}^{\infty} \frac{(-ig)^s}{s!} \theta(\lambda_1) \cdots \theta(\lambda_s) \int d\xi \int dR_s \, \Gamma_1 \cdots \Gamma_s$$

$$\times [x_1 \cdots x_{2N_0} \xi_1^{(1)} \cdots \xi_1^{(4)} \cdots \xi_1^{(1)} \cdots \xi_s^{(4)} \xi\xi\xi\xi]. \quad (9.26)$$

In the computation of (9.26) we made use of Eqs. (3.11) and (3.12). It is easy to see that Eq. (9.26) can be written

$$\delta_f \bar{K}_{2N_0} = (N_0 R_2^{(2)} + R_0 + R_2^{(1)} \Omega \int [\xi\xi_1] \hat{\delta}(\xi - \xi_1) \, d\xi_1) \bar{K}_{2N_0}$$

$$+ 2i R_2^{(1)} \frac{\partial \bar{K}_{2N_0}}{\partial m} + (iR_4 - 2\bar{g}R_2^{(2)}) \frac{\partial \bar{K}_{2N_0}}{\partial \bar{g}}. \quad (9.27)$$

In (9.27) derivatives are to be interpreted as formal term-by-term derivatives of expansion (9.12); comparing Eq. (9.27) with Eq. (9.11), we have

$$R_2^{(2)} = Z^{-1} \left\{ \delta_f Z + \frac{1}{Q} \frac{\partial Z}{\partial \overline{m}} \left(\frac{\partial \overline{\overline{m}}}{\partial \overline{g}} \delta_f \overline{\overline{g}} - \frac{\partial \overline{\overline{g}}}{\partial \overline{g}} \delta_f \overline{\overline{m}} \right) - \frac{1}{Q} \frac{\partial Z}{\partial \overline{g}} \right.$$

$$\left. \times \left(\frac{\partial \overline{\overline{m}}}{\partial \overline{m}} \delta_f \overline{\overline{g}} - \frac{\partial \overline{\overline{g}}}{\partial \overline{m}} \delta_f \overline{\overline{m}} \right) \right\} [R_0 + R_2^{(1)} \Omega \int [\xi \xi_1] \hat{\delta}(\xi - \xi_1) d\xi_1]$$

$$= A^{-1} \left\{ \delta_f A + \frac{1}{Q} \frac{\partial A}{\partial \overline{m}} \left(\frac{\partial \overline{\overline{m}}}{\partial \overline{g}} \delta_f \overline{\overline{g}} - \frac{\partial \overline{\overline{g}}}{\partial \overline{g}} \delta_f \overline{\overline{m}} \right) + \right.$$

$$\left. - \frac{1}{Q} \frac{\partial A}{\partial \overline{g}} \left(\frac{\partial \overline{\overline{m}}}{\partial \overline{m}} \delta_f \overline{\overline{g}} - \frac{\partial \overline{\overline{g}}}{\partial \overline{m}} \delta_f \overline{\overline{m}} \right) \right\}, \tag{9.28}$$

$$2i R_2^{(1)} = - \frac{1}{Q} \left(\frac{\partial \overline{\overline{m}}}{\partial \overline{g}} \delta_f \overline{\overline{g}} - \frac{\partial \overline{\overline{g}}}{\partial \overline{g}} \delta_f \overline{\overline{m}} \right),$$

$$i R_4 - 2\overline{g} R_2^{(2)} = \frac{1}{Q} \left(\frac{\partial \overline{\overline{m}}}{\partial \overline{m}} \delta_f \overline{\overline{g}} - \frac{\partial \overline{\overline{g}}}{\partial \overline{m}} \delta_f \overline{\overline{m}} \right).$$

In conclusion, starting with the perturbative expansion of K_{2N_0} we have shown that the $\hat{\delta}$ procedure is a renormalization. We would arrive at the same result by starting with equations of the first or second type and using the procedure adopted in Chapter 8 for the $g\phi^3$ theory.

4. A BRIEF MENTION OF THE RENORMALIZATION GROUP

At this point mention of the "renormalization group" is in order; this is defined as the group of all transformations that leave invariant (but for inessential terms) the propagators and therefore the physical content of the theory. Equations (9.1) and (9.2) are therefore satisfied.

These transformations form a group with a multiplication defined in the following manner. Given three regularizations $\int$, $\int'$, and $\int''$, we define the element $\gamma(\int, \int'')$ (transforming $\int$ into $\int''$) as the product of the element $\alpha(\int, \int')$ (transforming $\int$ into $\int'$) and the element $\beta(\int', \int'')$ (transforming $\int'$ into $\int''$):

$$\alpha(\int \int') \beta(\int' \int'') = \gamma(\int \int''). \tag{9.29}$$

It is clear that all formal properties of groups are easily inferred from this definition of product. *What* group cannot, of course, be known unless specific information is given as to how each renormalization is performed. A relevant amount of research on this subject may still be demanded (see the considerations at the end of this chapter). The connection between any two renormalizations was derived by us without difficulty via the finite method, with group-theoretical means; only conclusions and references need be indicated here: it all amounts to demonstrating that a formula like (9.1) holds (except for inessential terms, such as the standard additional gauge contributions to this formula in electrodynamics, which are of no physical consequence). Thus for any $\int$, $\int'$

$$K_{2N_0}(g, m^2, \textstyle\int) = A(g, m^2, \textstyle\int, \int')Z^{N_0}(g, m^2, \textstyle\int, \int')$$
$$\times K_{2N_0}(g'(g, m^2, \textstyle\int), m'^2(g, m^2, \textstyle\int), \int'). \qquad (9.30)$$

It is instructive to exhibit the connection between the finite and the infinitesimal method of proving that a regularization is a renormalization. The first obvious remark is that the latter can also be regarded as a prescription for actually performing an infinitesimal change of renormalization, and its operations are therefore infinitesimal generators of the finite group (9.29). The transformation between any two given different renormalizations can thus be thought of also as accomplished through an infinite chain of infinitesimal transformations; this is seen, quite simply, as follows.

Denote with $\int$ and $\int'$ two renormalizations (e.g., the $\hat{\delta}$ procedure and the σ or any other standard normalization procedure, or the two $\hat{\delta}$ procedures obtained by choosing two different functions f_1 and f_2) and introduce a new renormalization [59]

$$\textstyle\int_\alpha = \alpha \int + (1 - \alpha) \int', \qquad 0 \leqslant \alpha \leqslant 1. \qquad (9.31)$$

The family of renormalizations $\int_\alpha$ realizes, with the metric

$$d(\textstyle\int_\alpha, \int_\beta) = |\alpha - \beta|,$$

the continuous passage from $\int$ to $\int'$ when α goes from 0 to 1. Thus we have obtained the connection between $\int$ and $\int'$ through successive infinitesimal steps from $\int_\alpha$ to $\int_{\alpha + d\alpha}$. We fall thus into a familar situation: through a sequence of locally well-defined infinitesimal operations, a finite transformation is accomplished by a parameter integration over a finite domain. But at this point, a major question often arises: *Under what conditions at large is the transition from local to global legitimate?* We may well expect (and we shall find in the simple models of the next chapter: whatever their physical value, their mathematics is not objectionable) to discover, handling the matter at large and globally, *disconnected domains,* such that (1) transitions among points of different domains are forbidden; (2) each domain is an equally legitimate candidate for the correct assessment of mass and charge values.

This amounts to saying that *inequivalent representations* are, unless proved otherwise, to be expected a priori as the consequence of a consistent study of renormalization, in any study "at large." It is encouraging to find that the concept of inequivalent representation arises so naturally also in a field theory based on propagators rather than on formal axioms: It seems, so to speak, to force its way into it, and has an immediate and intuitive mathematical explication. The difficulty ahead is, of course, to know to what extent results that may be obtained in this way are in fact model dependent.

5. A COMMENT ON NOTATION

This brief notational remark is intended for the reader who may still be interested in reading some of our past work on the subject. Chronologically, our approach has been just about the reverse of the one followed here; the methodology adopted by us required a clear-cut distinction between combinatoric and analytic properties; the first being, as explained in the Preface, an essential preliminary to further study. We thus took for granted what appeared obvious (perhaps only) to our intuition, and neglected at the beginning to state specific rules for performing regularizations or renormalizations.

Chronologically, therefore, our notation has followed this sequence:

$$
\int \cdots \int d\xi_1 \cdots d\xi_n \overset{\rightrightarrows}{\equiv} \int d\xi_1 \int_{\xi_1} d\xi_2 \int_{\xi_1\xi_2} d\xi_3 \cdots \int_{\xi_1\cdots\xi_{n-1}} d\xi_n
$$
$$
\equiv [\text{integral renormalized (or distribution}
$$
$$
\text{defined) by a given specific rule (such as the}
$$
$$
\sigma \text{ or } \hat{\delta} \text{ procedures described in the text)}].
$$
$$
(9.32)
$$

On the other hand, in order to meet the demands often made on us, we have chosen to follow the opposite direction in this book, thus reversing the arrows from right to left. This remark should suffice to supply the necessary link between the present notation and that of our earlier references: whenever combinatorics only is required (and in any case after some familiarity with this formalism is acquired), the short notation (at the left-hand side of (9.32)) is as adequate as, and handier than, the explicit one in which the rule is actually explained. We began it all in this way.

We must of course also apologize for erroneous, useless, or ill-proved statements that have marred some of our papers on the subject; these cover an arch of about 20 years, not certainly enough to become wise, but many times sufficient to get wiser.

Chapter 10

Self-Consistent Approximations and Mass Equations

1. INTRODUCTION

Thus far we have been concerned only with the study of the renormalization of the *equations* satisfied by Green's functions, which makes them meaningful from a distribution-theoretical point of view (perturbative expansions come out then automatically renormalized).

The scope and depth of our perspective can be increased by considering, in this final chapter, some concrete models; these are conveniently obtained by taking the simplest self-consistent approximations to standard field theories. Peculiar features emerge, which cannot be predicted from perturbative expansions but are fully accounted for in the global framework outlined in the preceding chapter.

In all the examples reported in this chapter we find, as the crucial stop, a nonlinear mass equation, which admits, as a rule, of more than one solution. This is expected to be the general case. An interpretation of this fact, that is, of the inequivalent representations that appear in this way in our formalism, can be given without difficulty in terms of a spontaneous breakdown of symmetries [60]–[63]; the situation presents a substantial analogy to classical ones that arise in the study of phase transitions and other phenomena [64, 65].

It can also happen that the mass equation yields "ghosts" besides "particles": this case is treated as easily as (and leads, in the Lee model, to solutions that coincide with) the standard ones.

2. SOME HARTREE–FOCK APPROXIMATIONS

In any study of this nature it is essential to recall that truncation of a renormalizable theory, made in order to obtain a solvable approximation, or "model," *easily destroys its renormalizability*. This point is studied in

[66]; the considerations made there should be kept in mind in any attempt to extend the results reported later, which refer to very simple models, for which renormalizability remains obvious.

A. Hartree–Fock Approximation in the $g\phi^4$ Theory

This is expressed clearly by the first type equation [67, 68]

$$G(xy) = [xy] + 12igG(0) \int d\xi \, [x\xi] \, G(\xi y) \tag{10.1}$$

that is,

To avoid formal complications, we do not explicitly indicate, as a rule, regularization and FP integrals, which must, however, be understood to be applied *everywhere*: thus,

$$[xx] \equiv \mathrm{FP}[xx] \equiv \theta(\lambda) \int dy \, [xy] \, \hat{\delta}(x - y, \lambda) \tag{10.2}$$

$$\equiv \frac{m_0{}^2}{16\pi^2} \log \frac{m_0{}^2}{4M^2}$$

This model still gives a free particle; the free mass, however, is changed into the "physical" one m_f by its interaction with the average self-field. From now on, to facilitate comparison with the references, we shall write: $[xy] = G_0(x - y)$. The Lehmann representation [69]

$$G(x) = \int_0^\infty \rho(\alpha) G_{(0)}(x, \alpha) \, d\alpha \tag{10.3}$$

or

$$\hat{G}(P) = \int_0^\infty \rho(\alpha) \tilde{G}_{(0)}(p, \alpha) \, d\alpha$$

gives, with

$$\bar{\Sigma}(\alpha) = -12gG(0) = -12g \, \mathrm{FP} \int_0^\infty \rho(\alpha) G_{(0)}(0, \alpha) d\alpha, \tag{10.4}$$

the result

$$\rho(\alpha) = \lim_{\epsilon \to 0} \frac{1}{\pi} \frac{\epsilon - \mathrm{Im} \, \bar{\Sigma}(\alpha)}{[\alpha - m_0{}^2 - \mathrm{Re} \, \bar{\Sigma}(\alpha)]^2 + [-\epsilon + \mathrm{Im} \, \bar{\Sigma}(\alpha)]^2} \tag{10.5}$$

that is,

$$\rho(\alpha) = \delta(\alpha - m_0{}^2 + 12g \, \mathrm{FP} \int_0^\infty \rho(\alpha) G_{(0)}(0, \alpha) \, d\alpha) = \delta(\alpha - m_f{}^2) \tag{10.6}$$

where, keeping in mind (10.2),

$$m_f^2 = m_0^2 - \frac{3}{4\pi^2} g m_f^2 \log \frac{m_f^2}{4M^2} \tag{10.7}$$

Equation (10.7) is a *mass equation,* which is more conveniently written, with $\bar{m}_0 = m_0/2M$, $\bar{m}_f = m_f/2M$, as

$$\bar{m}_f^2 = \bar{m}_0^2 - \frac{3}{4\pi^2} g \bar{m}_f^2 \log \bar{m}_f^2. \tag{10.8}$$

For given $g < 0$ and $\bar{m}_0$, $\bar{m}_f$ can have in general *two* values (for a detailed discussion see [67]). When g vanishes, the value of m_f that would come also from perturbation theory tends to $\bar{m}_0$, the other to infinity. The model contains no scattering, so both masses are stable; *either* one can occur, not *both* simultaneously. The obvious way of dealing with such situation is to consider isotopic multiplets. We show next the behavior of an N-multiplet.

B. Hartree–Fock Approximation with $\mathscr{L} = g\left(\sum\limits_{i=1}^{N} \phi_i^2(x)\right)^2$

If we have N fields, with the *same* bare mass m_0, coupled through the quartic interaction

$$\mathscr{L} = g\left(\sum_{i=1}^{n} \phi_i^2(x)\right)^2, \tag{10.9}$$

the same model we used before now gives the equations

$$G^{(i)}(x-y) = G_{(0)}^{\{i\}}(x-y) + 12ig \int G_{(0)}^{\{i\}}(x-\xi)G_{(0)}^{(i)}G^{(i)}(\xi-y)\,d\xi$$
$$+ \sum_{\substack{i \neq j}}^{1,\ldots,N} 4ig \int G_{(0)}^{\{i\}}(x-\xi)G^{(j)}(0)G^{(i)}(\xi-y)\,d\xi \tag{10.10}$$

Hence the mass equations for the physical masses $m_f^{(i)}$ [70]:

$$(m_f^{(i)})^2 = m_0^2 - \frac{g}{4\pi^2}\left\{3(m_f^{(i)})^2 \log\left(\frac{m_f^{(i)}}{M}\right)^2\right.$$
$$\left. + \sum_{\substack{j \neq i}}^{1,\ldots,N} (m_f^{(j)})^2 \log\left(\frac{m_f^{(j)}}{M}\right)^2\right\}. \tag{10.11}$$

Call now

$$x_i = \left(\frac{m_f^{(i)}}{M}\right)^2, \qquad \bar{m}_0 = \frac{m_0}{M},$$

so that (10.11) read

$$x_i + \frac{g}{2\pi^2} x_i \log x_i = \overline{m}_0{}^2 - \frac{g}{4\pi^2} \sum_{j=1}^{N} x_i \log x_i, \tag{10.12}$$

which is easily transformed into

$$x_i + \frac{g}{2\pi^2} x_i \log x_i = \frac{1}{N+2} \left[2\overline{m}_0{}^2 + \sum_{j=1}^{N} x_j \right]. \tag{10.13}$$

It is clear from (10.13) that, no matter what set of solutions is chosen, the physical masses of that N-plet can have at most two distinct values.

3. PHYSICAL INTERPRETATION

A possible interpretation of the previous results is to consider the model as describing an N-multiplet of identical particles with a Lagrangian invariant under isospace rotations. The mass equation permits us to describe two different situations:

(i) *symmetrical*—all masses have the same value m_{f_1} (or m_{f_2});
(ii) *with symmetry breakdown*—K masses have values m_{f_1}, $N - K$ have values m_{f_2}.

With regard to case (ii), it is to be remarked that the Goldstone particle [60] cannot appear at this level, because the symmetry-breaking terms are quadratic [61] (whereas in the σ model they are linear [62, 63]); the Goldstone particle will appear as a bound state (as it can in fact be shown) in analogy with the Nambu–Jona Lasinio model [64].

It is also relevant to note that, in case (ii), the *renormalization parameters are no longer inessential.* It could not be otherwise with inequivalent representations. This point certainly deserves deeper discussion, but this is better deferred until more realistic examples have been sufficiently studied.

4. HARTREE-FOCK APPROXIMATION FOR NONPOLYNOMIAL LAGRANGIANS

Consider the Lagrangian [71]

$$\mathscr{L}(x) = -ig \sum_{n=0}^{\infty} \frac{a_n}{n!} \phi^n(x) = -igl[\phi(x)]. \tag{10.14}$$

The equations for Green's functions are written most easily (we recall that the normal product is *not* used, and the point loops are evaluated as before) by introducing the notation

$$K(x_1 \cdots x_r \underbrace{y \cdots y}_{m} x_s \cdots x_n) = K(x_1 \cdots x_r \, y^m x_s \cdots x_n),$$

and if

$$Z(y) = \Sigma \, c_n \, y^n,$$

then

$$K(x_1 \cdots x_r \, Z(y) \, x_s \cdots x_n) = \Sigma \, c_n K(x_1 \cdots x_r \, y^n \, x_s \cdots x_n).$$

We find then

$$G(x_1 - x_2) = G_0(x_1 - x_2) + g \int G_{(0)}(x_1 - y) G\left(x_2, \frac{\partial l(y)}{\partial y}\right) dy, \tag{10.15}$$

and so on (see [71]). Taking for simplicity's sake $a_{2n+1} = 0$, we obtain a Hartree–Fock approximation by setting

$$G(x_1, \ldots, x_{2m}) = \underset{\text{all pairs}}{\Sigma} \, G(x_{i_1} - x_{i_2}) \cdots G(x_{2m-1} - x_{2m}); \tag{10.16}$$

that is, equal to the hafnian made out of two-point free propagators with *physical* mass m_f. Then

$$G(x_2 y^{2m-1}) = (2m - 1)!! \, [\text{FP } G(0)]^{m-1} G(x_2 y) \tag{10.17}$$

and (10.15) becomes

$$G(x_1 - x_2) = G_{(0)}(x_1 - x_2) + \frac{g}{2} \sum_{m=0}^{\infty} \frac{a_{2m+2}}{(m+1)!} \left[\frac{G(0)}{2}\right]^m$$
$$\cdot \int G_{(0)}(x_1 - y) G(y - x_2) \, dy, \tag{10-18}$$

which gives the mass equation

$$m_f^2 = m_0^2 - \frac{g}{2} \sum_{m=0}^{\infty} \frac{a_{2m+2}}{(m+1)!} \left[\frac{m_f^2}{32\pi^2} \log \frac{m_f^2}{M^2}\right]^m. \tag{10.19}$$

At the level required by this approximation, all steps concerning products of distributions can be easily justified. The solutions of (10.19), as an equation in m_f, will depend of course on the choice of coefficients a_{2m}.

5. THE LEE MODEL

The Lee model [72] $V \rightleftarrows N + \theta$ is an interesting test case, from the points of view both of renormalization and of the "mass equation"; since it can be solved exactly, the discussion can be complete and final. Its first sector can of course be studied apart from the others [73]; this will suffice for our purposes. The Green function of the V particle clearly satisfies the equation

$$G^V(x - y) = G_{(0)}{}^V(x - y)$$

$$- g^2 \int G_{(0)}{}^V(x - \xi)G_{(0)}{}^N(\xi - \eta)G_{(0)}{}^\theta(\xi - \eta)G^V(\eta - y) \, d\xi \, d\eta$$

$$(10.20)$$

where

$$G_{(0)}^{V,N}(x - y) = \theta(x^0 - y^0)\delta^3(\mathbf{x} - \mathbf{y}) \exp[-im_0^{V,N}(x_0 - y_0)] \quad (10.21)$$

and $G_{(0)}{}^\theta(x - y)$ is the free propagation for a boson of mass μ (g is the coupling constant).

Renormalization will be done here, to agree with [73], with the σ procedure; that is,

$$\int G_{(0)}{}^V(x - \xi)G_{(0)}{}^N(\xi - \eta)G_{(0)}{}^\theta(\xi - \eta)G^V(\eta - y) \, d\xi \, d\eta$$

$$= \theta(\lambda) \int G_{(0)}{}^V(x - \xi)G_{(0)}{}^N(\xi - \eta)G_{(0)}{}^\theta(\xi - \eta)G^V(\eta - y)f(\lambda)$$

$$(\xi - \eta)^{2\lambda} \, d\xi \, d\eta. \quad (10.22)$$

Only the first two terms of the Taylor expansion

$$f(\lambda) = 1 + \alpha\lambda + \cdots$$

actually contribute in (10.22).

The computation, which is easily done, is reported in [73]. The renormalized Green function in momentum space is given by

$$G(K_0) = \frac{1}{K_0 - m_0{}^V + \Sigma(K_0)} = \int\limits_0^\infty \frac{\rho(\tau)}{K_0 - \tau} \, d\tau \quad (10.23)$$

where

$$\rho(\tau) = Z_1\delta(\tau - \bar{m}_1{}^V) + Z_2\delta(\tau - \bar{m}_2{}^V) + \theta[\tau - (m^N + \mu)]\sigma(\tau); \quad (10.24)$$

$\overline{m}_1{}^V$ and $\overline{m}_2{}^V$ are the two possible solutions of the equation

$$\overline{m}^V - m_0{}^V + \Sigma(\overline{m}^V) = 0; \tag{10.25}$$

and

$$Z_{1,2} = \left[1 + \left(\frac{\partial\Sigma}{\partial K_0}\right)_{K_0 = \overline{m}_{1,2}^V}\right]^{-1}. \tag{10.26}$$

The computation made in [73] shows that

1. $\sigma(\tau) > 0$ in (10.24), whatever the arbitrary constant α in $f(\lambda)$;
2. Eq. (10.25) has *two* solutions, both *simultaneously* existing;
3. one of the two values Z_1 and Z_2 from (10.26) that correspond to these two solutions is *positive* (i.e., the real particle); the other is *negative* (i.e., the ghost).

In this case the expected standard results appear quite simply, as solutions of the mass equation.

References

1. E. R. Caianiello, *Nuovo Cimento* 10, 1634 (1953); 11, 492 (1954).
2. See, e.g., T. Muir, *Contribution to the History of Determinants*, 5 vols. (the last published in London, 1930).
3. E. Cartan, *Les systèmes différentiels extérieurs et leurs applications géométriques*, Hermann, Paris, 1971.
4. C. Chevalley, *Theory of Lie Groups*, Princeton Univ. Press, 1957.
5. Cf. D. Littlewood, *The Theory of Group Characters*, Oxford Univ. Press, 1950.
6. N. G. de Bruijn, *J. Indian Math. Soc.* 132, XIX (1955).
7. H. S. Green and C. A. Hurst, *Order–Disorder Phenomena*, Wiley (Interscience), New York, 1964.
8. E. W. Montroll, *J. Math. Phys.* 4, 308 (1963).
9. P. W. Kasteleyn, *Physica* 27, 1209 (1961); *J. Math. Phys.* 4, 287 (1963).
10. E. R. Caianiello and S. Fubini, *Nuovo Cimento* 9, 1218 (1952).
11. N. S. Goel, S. C. Maitra, and E. W. Montroll, "On the Volterra and other models of population of interacting biological species," *Rev. Mod. Phys.* (in press).
12. J. Schwinger, *Phys. Rev.* 93, 615 (1954).
13. E. R. Caianiello and F. Guerra, unpublished notes.
14. E. R. Caianiello, *La Ricerca* VII, No. 1 (1956).
15. E. R. Caianiello, *Nuovo Cimento* 13, 637 (1959); 14, 185 (1959).
16. E. R. Caianiello, F. Guerra, and M. Marinaro, *Nuovo Cimento* 60A, 713 (1969).
17. N. N. Bogolubov and D. V. Shirkov, *Introduction to the Theory of Quantized Fields*, Wiley (Interscience), New York, 1959.
18. J. M. Jauch and F. Rohrlich, *Theory of Photons and Electrons*, Addison-Wesley, Reading, Mass., 1955.
19. J. D. Bjorken and S. D. Drell, *Relativistic Quantum Fields*, McGraw-Hill, New York, 1965.
20. K. O. Friedrichs, *Mathematical Aspects of the Quantum Theory of Fields*, Wiley (Interscience), New York, 1953.
21. R. F. Streater and A. S. Wightman, *PCT, Spin, Statistics and All That*, Benjamin, New York, 1964.
22. F. A. Berezin, *The Method of Second Quantization*, Academic Press, New York, 1966.
23. E. R. Caianiello, H. Umezawa, and B. Preziosi, *Nuovo Cimento* 20, 1001 (1961).
24. E. R. Caianiello and K. Y. Shen, *Nuovo Cimento* 20, 1038 (1961).
25. E. R. Caianiello, *Nuovo Cimento* 9, 155 (1955).
26. A. Buccafurri and E. R. Caianiello, *Nuovo Cimento* 8, 170 (1958).
27. H. Lehmann, K. Symanzik, and W. Zimmermann *Nuovo Cimento* 1, 205 (1955); 6, 319 (1957).

28. A. Campolattaro and M. Marinaro, *Nuovo Cimento* 22, 879 (1961).
29. G. Guerra and M. Marinaro, *Nuovo Cimento* 42, 306 (1966).
30. D. J. Candlin, *Nuovo Cimento* 12, 380 (1954).
31. K. A. Brueckner and J. L. Garmel, *Phys. Rev.* 109, 1023 (1958); J. Goldstone, *Proc. Roy. Soc. (London)* A293, 267 (1957).
32. E. R. Caianiello, *Nuovo Cimento* 12, 561 (1954).
33. E. R. Caianiello and A. Campolattaro, *Nuovo Cimento* 26, 390 (1962).
34. E. R. Caianiello, A. Campolattaro, and M. Marinaro, *Nuovo Cimento* 38, 1777 (1965).
35. F. G. Tricomi, *Funzioni Ipergeometriche Confluenti,* Rome (1954).
36. T. Carleman, *Les Functions Quasi-analitiques,* Paris (1926).
37. L. D. Landau, A. A. Abrikosov, and I. M. Khalatnikov, *Dokl. Akad. Nauk SSR* 95, 773 (1954).
38. B. Zumino, Univ. of Calif. Rept. UCRN 8896 (1959).
39. H. M. Fried, *Phys. Rev.* 115, 220 (1959).
40. S. Okubo, *Nuovo Cimento* 15, 949 (1960).
41. F. Bloch and A. Nordsieck, *Phys. Rev.* 52, 54 (1937).
42. D. R. Yennie and H. Suura, *Phys. Rev.* 105, 1378 (1957).
43. E. R. Caianiello and S. Okubo, *Nuovo Cimento* 17, 355 (1960).
44. I. M. Gel'fand and G. E. Shilov, *Generalized Functions,* Academic Press, New York, 1964.
45. O. S. Parasiuk, *Izv. Akad. Nauk Ser. Mat.* 20, 843 (1956); N. N. Bogoliubov and O. S. Parasiuk, *Acta Math.* 97, 227 (1957); O. S. Parasiuk, *Ukrainskii Math. Z.* 12, 287 (1960).
46. M. Riesz, *Acta Math.* 81 (1949).
47. C. G. Bollini, J. J. Giambiagi, and A. Gonzalez Dominguez, *Nuovo Cimento* 31, 550 (1964); B. Preziosi, *Nuovo Cimento* 31, 187 (1964).
48. W. Guttinger, *Fortschr. Phys.* 14, 489 (1966).
49. G. Guerra and M. Marinaro, *Nuovo Cimento* 60A, 756 (1969).
50. E. R. Speer, *J. Math. Phys.* 9, 1404 (1968).
51. F. Guerra, *Nuovo Cimento* 1A, 523 (1971).
52. M. Marinaro, "Comparison between different approaches to renormalization," *Nuovo Cimento* (in press).
53. E. R. Caianiello and A. Campolattaro, *Nuovo Cimento* 20, 422 (1961).
54. F. Guerra and M. Marinaro, *Nuovo Cimento* 42, 285 (1966).
55. F. J. Dyson, *Phys. Rev.* 75, 486, 1736 (1949).
56. A. Salam, *Phys. Rev.* 82, 217 (1951).
57. P. T. Matthews and A. Salam, *Rev. Mod. Phys.* 23, 311 (1951).
58. E. R. Caianiello and M. Marinaro, *Nuovo Cimento* 27, 1185 (1962).
59. F. Esposito, U. Esposito, and F. Guerra, *Nuovo Cimento* 60A, 772 (1969).
60. J. Goldstone, A. Salam, and S. Weinberg, *Phys. Rev.* 127, 965 (1962).
61. K. Symanzik, DESY 70/62 November 1970.
62. B. W. Lee, *Nucl. Phys.* B2, 649 (1969).
63. M. Gell-Mann and M. Levy, *Nuovo Cimento* 16, 705 (1960).
64. Y. Nambu and G. Iona-Lasinio, *Phys. Rev.* 122, 345 (1961); 124, 246 (1961).
65. W. Heisenberg, *Introduction to Unified Field Theory of Elementary Particles,* New York, 1966.
66. M. Marinaro and G. Scarpetta, *Nuovo Cimento* 67B, 204 (1970).
67. E. R. Caianiello, M. Marinaro, and G. Scarpetta, *Nuovo Cimento* 3A, 195 (1971).
68. Application of this technique to the anharmonic oscillator gives the energy gap

between ground and first excited state to within 5%: A. Melkman, "Approximating the Coupled Equation of Field Theory," Ph.D. thesis, Univ. of California, Berkeley (1971).

69. H. Lehmann, *Nuovo Cimento* 11, 342 (1954).
70. E. R. Caianiello, "Analytic Renormalization and Mass Equations," report presented at Congress on Renormalization Theory, Marseilles, June 1971.
71. E. R. Caianiello and M. Marinaro, *Lett. Nuovo Cimento* 1, 899 (1971).
72. T. D. Lee, *Phys. Rev.* 95, 1329 (1954).
73. F. Esposito and U. Esposito, *Nuovo Cimento* 6A, 277 (1971).

Subject Index